NMR Spectroscopy in Inorganic Chemistry

NMR Spectroscopy in Inorganic Chemistry

SECOND EDITION

Jonathan A. Iggo

Reader in Inorganic Chemistry
Department of Chemistry, University of Liverpool

Konstantin V. Luzyanin

Lecturer in Analytical Sciences
Department of Chemistry, University of Liverpool

OXFORD
UNIVERSITY PRESS

OXFORD

UNIVERSITY PRESS

Great Clarendon Street, Oxford, OX2 6DP,
United Kingdom

Oxford University Press is a department of the University of Oxford.
It furthers the University's objective of excellence in research, scholarship,
and education by publishing worldwide. Oxford is a registered trade mark of
Oxford University Press in the UK and in certain other countries

First Edition 2000

Impression: 1

Published in the United States of America by Oxford University Press
198 Madison Avenue, New York, NY 10016, United States of America

British Library Cataloguing in Publication Data
Data available

Library of Congress Control Number: 2019952139

ISBN 978-0-19-879485-1

Printed in Great Britain by
Bell & Bain Ltd., Glasgow

Preface to the First Edition

Most students first meet NMR spectroscopy in 'Organic' Chemistry. The approach usually adopted is to concentrate on interpretation of ^1H NMR spectra and ignore much if not all of the physical basis of the technique. Whilst this approach often enables the student to interpret simple ^1H spectra many students have difficulty in interpreting the NMR spectra of other nuclei. Some students even seem unaware that NMR extends beyond protons! The aim of this book is to provide a non-mathematical grounding in the 'physics' of NMR spectroscopy and then use this to explore the use of NMR spectroscopy in 'Inorganic' Chemistry. The Author hopes that chapters 1 'Fundamentals' and 2 'Structure Determination' will be useful and accessible to 2nd year chemistry students and above, whilst chapters 3 'Dynamic Processes' and 4 'The Solid State' are intended for 3rd year students and beyond.

Perhaps the most widespread use of NMR spectroscopy in inorganic chemistry is in the structural characterization of organometallic, of metal carbonyl cluster, and of boron cluster compounds. This in part reflects the huge amount of synthetic work done in these areas and in part the fact that there is a large number of spin ½ nuclei available for study in these areas, of which the most important are ^{13}C, and ^{31}P. The student should not forget, however, that there are other equally important nuclei, e.g. ^1H, ^2D, ^{11}B, ^{15}N, ^{17}O, and ^{119}Sn, for which NMR spectra can be obtained, albeit (with the exception of ^1H) with greater difficulty. The examples used in this book inevitably reflect the amount of work done with each nucleus, although the author has tried to include examples from many different areas where suitable examples are available. Not all the work described is the author's own; where possible acknowledgement is given in the text for specific examples from others' work. Where data has been gleaned from a variety of sources, e.g. to provide comparative data, this has not been possible, and acknowledgement to those workers is made here.

Preface to the Second Edition

In the twenty years since the first edition of this book 'inorganic' NMR has found application in an increasingly diverse range of fields from biomedical, to food, to materials science. In preparing the second edition we have attempted to reflect this diversity, incorporating many new examples to illustrate these applications as well as new examples from more traditional molecular, inorganic chemistry.

We have completely revised the text while retaining the non-mathematical, example-based approach of the first edition. Following the advice from reviewers, we have sub-divided the introductory material into separate chapters. 'Fundamentals', 'Structure determination', and 'Factors influencing the chemical shift and coupling constant' now each receive chapters in their own right. A workflow for the interpretation of spectra that avoids the 'pattern-based' and single-spin approaches is used and a discussion of situations where ambiguities can arise and how these can be disentangled is presented. A new Chapter 4 dealing with 'Experimental methods: pulses, the vector model, and relaxation' has been added before a completely revised and expanded Chapter 5 'Polarization transfer and 2-D NMR spectroscopy' methods. Chapters 6 on 'Dynamic NMR spectroscopy' and 7 on 'The solid state' have been expanded, with new examples added. Exercises have been added at the end of each chapter and detailed solutions are available in the Online Resources.

We thank Kelvin Tan and Leong Weng Kee for original versions of some figures and Jianliang Xiao and Naomi Ritchie for permission to include some unpublished work. We hope readers will find the revisions helpful.

Jonathan A. Iggo
Konstantin V. Luzyanin
Liverpool, July 2019

Contents

1 Fundamentals

1.1 Introduction

Nuclear magnetic resonance (NMR) spectroscopy is commonly first encountered in organic chemistry, and attention is focused on the NMR spectroscopy of a single element, hydrogen. There are, however, many other elements that have **isotopes** with **nuclear spin**; if these elements are taken into account, NMR becomes perhaps the most important spectroscopic technique today for the characterization of inorganic compounds in solution and is of growing importance in the solid state. In solution, spin ½ nuclei give well-resolved, sharp resonances for each NMR active site in the molecule whilst, in the solid state, cross-polarization has increased the sensitivity and **magic angle spinning** the resolution possible.

Although the presence of several different elements possessing nuclear spin in a molecule might at first seem likely to complicate the interpretation of the NMR spectrum, in fact it offers the possibility of simplification, since each nuclide/element has a characteristic resonance frequency so we can record a series of NMR spectra, looking at each nuclide in turn. The chemical shifts and coupling patterns observed will, of course, depend on the other nuclides present. The aim of this book is to introduce the reader to the NMR spectroscopy of the 'inorganic' elements and to its application in structure determination and the characterization of reaction processes in inorganic chemistry.

The rules for interpreting NMR spectra are independent of the particular elements present, the spectrum depending only on the nuclear spins present (Figure 1.1). Therefore, it is sensible first to focus on the nuclear spins present and the way these behave rather than the elements present. By developing an understanding of why NMR spectra look the way they do based on the spins, rather than the elements present, we can apply our knowledge to the interpretation of any NMR spectrum, not just that of protons. The magnitude of the chemical shift and of the coupling constants will, of course, vary widely depending on the nuclides under study. For example, rhodium chemical shifts span a range +/−4000 ppm and platinum–phosphorus coupling constants are measured in thousands(!) of hertz rather than the familiar proton ranges of 0–10 ppm and 0–20 Hz, respectively.

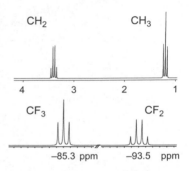

Figure 1.1 The rules for interpreting NMR spectra depend only on the nuclear spins present. Consider the ^{1}H NMR spectrum (top) of the ethyl group in CH_3CH_2OD and the ^{19}F NMR spectrum of the perfluoroethyl group in $CF_3CF_2OClO_3$ (bottom). Both ^{1}H and ^{19}F are spin ½ nuclei, in both compounds there is a group of three spins on one carbon and a group of two on the adjacent C. The spectra show the same coupling patterns because the same sorts of groups of spins are present, a group of three spin ½ and a neighbouring group of two spin ½. The coupling constants and chemical shifts differ.

Source: data taken from C. J. Schack and K. O. Christe (1979), *Inorg. Chem.* **18**, 2619. Note the 'old-fashioned' use of upfield shifts positive has been changed to the more usual downfield shifts positive in the figure.

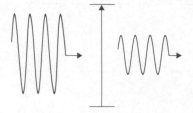

Figure 1.2 In a resonance spectroscopy transitions are excited and the frequency dependence of the absorption (or emission) of energy monitored to produce a spectrum

This chapter presents some basic theory of NMR and a formalism for the *prediction* of NMR spectra in solution. Chapter 2 discusses the *interpretation* of NMR spectra to elucidate the structure of an inorganic compound using the formalism presented in Chapter 1 rather than the widely followed 'pattern recognition' approach. Chapter 3 reviews the factors affecting the chemical shift and coupling constants in inorganic compounds. Chapters 4 and 5 discuss experimental methods and 2–D multiple-pulse techniques to edit and enhance the NMR spectrum to reveal specific information such as connectivities and molecular conformations. Chapter 6 introduces the NMR of dynamic systems and Chapter 7 that of solids.

1.2 Quantum numbers, energy levels, and things

NMR spectroscopy, as its name implies, is a resonance spectroscopy (Figure 1.2); we excite transitions between nuclear spin energy levels of the system, then monitor the resulting absorption or emission of energy due to the transitions in order to gain some knowledge about the system. Questions we would like answered include: 'How many levels are there?'; 'What are the allowed transitions between the energy levels?'; 'What factors influence the energy level separation?'; and 'How can we use this information to tell us about the chemistry of the sample?' It is clearly important that we have an understanding of the energy levels involved. The energy levels involved in NMR spectroscopy arise from the interaction of the nuclear spin with the spectrometer magnetic field, with the magnetic fields set up by other nuclear spins, and with the magnetic fields set up by the electrons in the sample. We will discover that the energy level separation is determined by the chemical environment (chemical shift) whilst the number of energy levels depends inter alia on interactions between spins (spin–spin coupling).

Nuclear spin

The nuclear spin arises from the unpaired proton and neutron spins in the nucleus in the same way as unpaired electrons give an atom an electronic spin; all isotopes with an odd atomic number and/or an odd atomic mass will therefore have a nuclear spin (Table 1.1). Sixty-three elements of the periodic table have at least one NMR active isotope; however, not all potential NMR active isotopes are useful since some only occur in paramagnetic compounds (most lanthanides) or have high quadrupole moments and broadening factors (e.g. ^{179}Hf, ^{191}Ir, and ^{197}Au). Although rules exist to predict the spin of a nuclide, it is easier to remember the spins of the common nuclei such as ^{1}H, ^{11}B, ^{13}C, ^{19}F, ^{31}P, and consult tables for the rest (Table 1.2).

Spin angular momentum and the nuclear magnetic moment

Any spinning body possesses angular momentum. In the case of a charged body, in NMR the nucleus, this spinning generates a magnetic moment; the

Table 1.1 All nuclides with an odd number of protons and/or an odd number of neutrons will have a nuclear spin. Spin zero nuclides are NMR inactive.

No. of protons	No. of neutrons	Spin
Even	Even	0
Even	Odd	½–integral
Odd	Even	½–integral
Odd	Odd	Integral

nuclear magnetic moment, μ. μ is a vector quantity and is proportional to the spin, I. The constant of proportionality γ is called the **gyromagnetic ratio** (equation 1.1). Every isotope has a different nuclear magnetic moment, μ, i.e., each isotope has its own, unique gyromagnetic ratio, γ. This is important since it means that each element/isotope will have a characteristic NMR resonance frequency; the resonances of one element will not appear in the NMR spectrum of another. However, the spins of different elements can couple to each other—heteronuclear coupling—which is important in many NMR experiments.

Eq. 1.1 The nuclear magnetic moment is proportional to the nuclear spin. The constant of proportionality, γ, is the gyromagnetic ratio.

$$\mu = \gamma I$$

Nuclear spin quantum numbers

The nuclear spin (strictly the nuclear spin angular momentum) is quantized and is given the quantum number I. I can take integral or half-integral values depending on the number of unpaired protons and neutrons (cf. the electron spin of an atom or ion depends on the number of unpaired electrons). For example, ^1H, ^{19}F, ^{31}P, and ^{103}Rh all have spin ½, ^2D and ^{14}N have spin 1, ^{11}B spin ³⁄₂, and ^{59}Co spin ⁵⁄₂. Nuclei that have spins greater than ½ are called **quadrupolar** and normally give broad NMR lines, although coupling to other nuclei may be well resolved. We will look first at spin ½ nuclei and extend our ideas to the NMR spectra of quadrupolar nuclei at the end of this chapter.

Quantum mechanically, the nuclear spin I and one of its components, I_x, I_y, I_z will be eigenfunctions of the nuclear spin Hamiltonian; we can know the total nuclear spin angular momentum and the magnitude of its component along one of the cartesian axes. Arbitrarily we choose to make I_z an eigenfunction and give it the quantum number m_I. m_I can take the values I, $(I-1)$, ...,$-I$ so there will be $2I+1$ levels associated with the nuclear spin I (Figure 1.3).

1.3 Nuclear spins interacting with a magnetic field

In the absence of an external magnetic field the m_I states are degenerate. This is reasonable; there is nothing with which the nuclear magnetic moment can interact so the orientation of the nuclear magnetic moment in space will not affect the energy of the system.

Effect of an applied field

If we place the nucleus in a magnetic field the different orientations of the nuclear magnetic moment will have different energies depending on the relative orientations of the nuclear magnetic moment and the applied field. Since the nuclear spin is quantized, only certain orientations of the nuclear magnetic moment with respect to the applied field are allowed; each orientation corresponds to a different m_I value so we get $2I+1$ discrete energy levels—the Zeeman splitting.

Figure 1.3 The number of m_I states (energy levels) for a nuclear spin I is $2I+1$. Here the two m_I states for a spin $I = ½$ nucleus are shown. In the absence of a magnetic field the m_I states all have the same energy. Nuclei with spins greater than ½ are discussed in section 1.8.

Figure 1.4 Placing the nucleus in a magnetic field, B_o, lifts the degeneracy of the m_I states giving $2I + 1$ different energy levels, each corresponding to one of the allowed orientations of the nuclear spin with respect to the applied field. Here the two m_I states for a spin $I = \frac{1}{2}$ nucleus are shown.

Spin ½ nuclei

Consider first the special case of a spin ½ nucleus. For spin ½ nuclei there are two energy levels corresponding to $m_I = +\frac{1}{2}$ and $-\frac{1}{2}$; there are two allowed orientations of the nuclear magnetic moment with respect to the applied field (Figure 1.4). We should be familiar with this picture from our study of ^1H NMR spectroscopy. Note, however, that the spins align at a fixed angle to the field. We will return to the significance of this later when we discuss quadrupolar nuclei at the end of this chapter and the vector model in Chapter 4.

Energies of the nuclear quantum states

Intuitively we would expect the energies of the allowed orientations of the nuclear magnetic moment to depend on the strength of the applied field, B_o, on the size of the nuclear magnetic moment, and on the orientation of the nuclear magnetic moment to the applied field, m_I (equation 1.2). This relationship can be derived either classically or quantum mechanically and readers are referred to the texts in the bibliography.

1.4 Nuclear magnetic resonance spectroscopy

Resonance frequencies

Eq. 1.2 The nuclear spin Hamiltonian can be solved to give the energies of the nuclear spin states. To first order, the energy levels for spin > ½ nuclei are evenly spaced.

$$E = m_I \gamma B \hbar$$

Eq. 1.3 Substitution in Plank's equation gives a typical NMR resonance frequency of 150 MHz.

$$E = h\nu$$
$$\nu \approx \frac{10^{-25}}{6.6 \times 10^{-34}}$$
$$\ldots \approx 150\,MHz$$

Typically in NMR spectroscopy the magnetic field, B_o, is of the order of 12 tesla and the gyromagnetic ratio, γ, is around 10^8 $T^{-1}s^{-1}$, giving a separation of adjacent nuclear spin energy levels of around 10^{-25} joules (equation 1.2). Substituting this value for the energy level spacing in Plank's equation gives a frequency of approximately 150 MHz (equation 1.3). Irradiation of the nuclear spins at this frequency will cause transitions between the nuclear spin energy levels. This frequency is known as the resonance or **Larmor frequency**. We are already familiar with the phenomenon of resonance, for example from IR and UV spectroscopies. We can now add NMR to our list of resonance spectroscopies.

Since each isotope has a distinct gyromagnetic ratio the nuclear spins of each element will resonate at characteristic, distinct frequencies. We will, therefore, not see resonances due to say ^{103}Rh in a ^{31}P NMR spectrum. The resonance frequencies and other relevant data for NMR active nuclei are given in Table 1.2.

Selection rules

Transitions are not allowed between all the energy levels. The selection rule for NMR is $\Delta mI = \pm 1$. Since the mI levels, in the absence of any other effects, are equally spaced (equation 1.2), a single resonance will be observed in the spectrum even for quadrupolar nuclei (for which there are more than two energy levels).

Table 1.2 Selected NMR properties of the elements. Common nuclei are given in bold, unobserved nuclei in italics.

Isotope	Spin	Natural abundancece (%)	Receptivity ^{13}C = 1.00	Larmor frequency[1] (MHz) at 11.74 T	Magnetogyric ratio (10^7 rad.s^{-1}.T^{-1})	Quadrupole Moment (fm^2)
^{1}H	1/2	99.9885	5870	500.13	26.7522208	
^{2}H	1	0.0115	0.007	76.773	4.10662919	0.286
^{107}Ag	1/2	51.839	0.205	20.244	−1.08918	
^{109}Ag	1/2	48.161	0.290	23.274	−1.2519	
^{27}Al	5/2	100	1220	130.318	6.976278	14.66
^{75}As	3/2	100	149	85.635	4.59615	31.4
^{197}Au	3/2	100	0.162	8.84	0.47306	54.7
^{10}B	3	19.9	23.2	53.732	2.87467955	8.459
^{11}B	3/2	80.1	777	160.462	8.584707	4.059
^{135}Ba	3/2	6.592	1.940	49.685	2.67769	16
^{137}Ba	3/2	11.232	4.620	55.579	2.99287	24.5
^{9}Be	3/2	100	81.5	70.277	−3.75966	5.288
^{209}Bi	9/2	100	848	80.367	4.3747	−51.6
^{79}Br	3/2	50.69	237	125.302	6.725619	30.5
^{81}Br	3/2	49.31	288	135.068	7.249779	25.4
^{13}C	1/2	1.07	1.00	125.758	6.728286	
^{43}Ca	7/2	0.135	0.051	33.659	−1.803069	−4.08
^{111}Cd	1/2	12.8	7.270	106.105	−5.698315	
^{113}Cd	1/2	12.22	7.940	110.995	−5.960917	
^{35}Cl	3/2	75.76	21.0	49.002	2.6241991	−8.165
^{37}Cl	3/2	24.24	3.88	40.789	2.1843688	−6.435
^{59}Co	7/2	100	1640	118.666	6.332	42 s
^{53}Cr	3/2	9.501	0.507	28.27	−1.51518	−15
^{133}Cs	7/2	100	284	65.598	3.53256	−0.343
^{63}Cu	3/2	69.15	382	132.612	7.111791	−22
^{65}Cu	3/2	30.85	208	142.055	7.6043	−20.4
^{161}Dy	5/2	18.889	0.526	17.19	−0.92	251
^{163}Dy	5/2	24.896	1.91	24.1	1.289	265
^{167}Er	7/2	22.869	0.677	14.42	−0.7716	357

Isotope	Spin	Natural abundance (%)	Receptivity ^{13}C = 1.00	Larmor frequency (MHz) at 11.74 T	Magnetogyric Ratio (10^7 rad.s^{-1}.T^{-1})	Quadrupole Moment (fm^2)
^{153}Eu	5/2	52.19	47.3	54.88	2.9357	241
^{19}F	1/2	100	4890	470.592	25.16233	
^{57}Fe	1/2	2.119	0.004	16.193	0.880627	
^{69}Ga	3/2	60.108	246	120.038	6.43886	17.1
^{71}Ga	3/2	39.892	335	152.523	8.18117	10.7
^{155}Gd	3/2	14.8	0.126	15.35	−0.8212	127
^{157}Gd	3/2	15.65	0.30	20.13	−1.077	135
^{73}Ge	9/2	7.76	0.644	17.446	−0.9360306	−19.6
^{3}He	1/2	0.000134	0.003	380.994	−20.3789473	
^{177}Hf	7/2	18.6	1.53	20.3	1.0858	337
^{179}Hf	9/2	13.62	0.438	12.75	−0.682	379
^{199}Hg	1/2	16.87	5.89	89.577	4.845793	
^{201}Hg	3/2	13.18	1.16	33.067	−1.78877	38.6
^{165}Ho	7/2	100	1160	105.71	5.654	358
^{127}I	5/2	100	560	100.063	5.38957	−71
^{113}In	9/2	4.29	88.5	109.357	5.8845	79.9
^{115}In	9/2	95.71	1990	109.592	5.8972	81
^{191}Ir	3/2	37.3	0.064	9	0.4812	81.6
^{193}Ir	3/2	62.7	0.137	9.77	0.5227	75.1
^{39}K	3/2	93.258	2.79	23.338	1.2500612	5.85
^{41}K	3/2	6.73	0.033	12.81	0.68614062	7.11
^{83}Kr	9/2	11.5	1.28	19.243	−1.033097	25.9
^{138}La	5	0.09	0.497	65.989	3.55724	45
^{139}La	7/2	99.91	356	70.647	3.808333	20
^{6}Li	1	7.59	3.79	73.6	3.937127	−0.0808
^{7}Li	3/2	92.41	1590	194.37	10.397704	−4.01
^{175}Lu	7/2	97.41	179	57.11	3.0547	349
^{176}Lu	7	2.59	6.05	40.53	2.168	497
^{25}Mg	5/2	10	1.58	30.616	−1.63884	19.94

(continued)

Table 1.2 Continued

Isotope	Spin	Natural abundance (%)	Receptivity ^{13}C = 1.00	Larmor frequency (MHz) at 11.74 T	Magnetogyric ratio (10^7 rad.s^{-1}.T^{-1})	Quadrupole Moment (fm²)
^{151}Eu	5/2	47.81	504	124.34	6.651	90.3
^{95}Mo	5/2	15.9	3.06	32.593	−1.7514	−2.2
^{97}Mo	5/2	9.56	1.96	33.277	−1.7884	25.5
^{14}N	1	99.636	5.90	36.141	1.9337798	2.044
^{15}N	1/2	0.364	0.022	50.697	−2.7126189	
^{23}Na	3/2	100	545	132.294	7.0808516	10.4
^{93}Nb	9/2	100	2870	122.413	6.5674	−32
^{143}Nd	7/2	12.2	2.43	27.25	−1.4574	−63
^{145}Nd	7/2	8.3	0.387	16.78	−0.898	−33
^{21}Ne	3/2	0.27	0.039	39.482	−2.113081	10.155
^{61}Ni	3/2	1.1399	0.24	44.692	−2.39477	16.2
^{17}O	5/2	0.038	0.065	67.8	−3.62806	−2.558
^{187}Os	1/2	1.96	0.001	11.415	0.6192897	
^{189}Os	3/2	16.15	2.32	38.837	2.10713	85.6
^{31}P	1/2	100	391	202.457	10.8394	
^{207}Pb	1/2	22.1	11.8	104.63	5.5767	
^{105}Pd	5/2	22.33	1.49	22.886	−1.23	66
^{141}Pr	5/2	100	1970	153.12	8.1907	−5.89
^{195}Pt	1/2	33.832	20.7	107.512	5.8385	
^{85}Rb	5/2	72.17	45	48.287	2.5927059	27.6
^{87}Rb	3/2	27.83	290	163.645	8.786403	13.35
^{185}Re	5/2	37.4	305	112.652	6.1057	218
^{187}Re	5/2	62.6	526	113.788	6.1682	207
^{103}Rh	1/2	100	0.186	15.936	−0.84677	
^{99}Ru	5/2	12.76	0.846	23.032	−1.228	7.9
^{101}Ru	5/2	17.06	1.58	25.814	−1.372	45.7
^{33}S	3/2	0.75	0.10	38.39	2.055685	−6.78
^{121}Sb	5/2	57.21	548	119.684	6.4435	−36
^{123}Sb	7/2	42.79	117	64.813	3.4892	−49

Isotope	Spin	Natural abundance (%)	Receptivity ^{13}C = 1.00	Larmor frequency (MHz) at 11.74 T	Magnetogyric Ratio (10^7 rad.s^{-1}.T^{-1})	Quadrupole Moment (fm²)
^{55}Mn	5/2	100	1050	123.978	6.64525453	33
^{45}Sc	7/2	100	1780	121.49	6.5088	−22
^{77}Se	1/2	7.63	3.15	95.382	5.125388	
^{29}Si	1/2	4.685	2.16	99.362	−5.31903	
^{147}Sm	7/2	14.99	1.34	20.84	−1.115	−25.9
^{149}Sm	7/2	13.82	0.692	17.18	−0.9192	7.5
^{115}Sn	1/2	0.34	0.711	163.636	−8.8013	
^{117}Sn	1/2	7.68	20.8	178.208	−9.5888	
^{119}Sn	1/2	8.59	26.6	186.502	−10.0317	
^{87}Sr	9/2	7	1.120	21.675	−1.163938	33.5
^{181}Ta	7/2	99.988	220	59.964	3.2438	317
^{159}Tb	3/2	100	408	120.22	6.431	143.2
^{123}Te	1/2	0.89	0.961	130.883	−7.059101	
^{125}Te	1/2	7.07	13.4	157.79	−8.510843	
^{47}Ti	5/2	7.44	0.918	28.195	−1.51054	30.2
^{49}Ti	7/2	5.41	1.20	28.203	−1.51095	24.7
^{203}Tl	1/2	29.52	340	285.69	15.539339	
^{205}Tl	1/2	70.48	836	288.494	15.692186	
^{169}Tm	1/2	100	3.32	41.37	−2.21	
^{235}U	7/2	0.7204	0.007	9.209	−0.52	493.6
^{50}V	6	0.25	0.818	49.865	2.67065	21
^{51}V	7/2	99.75	2250	131.549	7.0455139	−5.2
^{183}W	1/2	14.31	0.063	20.837	1.1282407	
^{129}Xe	1/2	26.4006	33.5	139.087	−7.4521	
^{131}Xe	3/2	21.2324	3.51	41.23	2.209077	−11.4
^{89}Y	1/2	100	0.70	24.507	−1.316279	
^{171}Yb	1/2	14.28	4.63	87.519	4.7288	
^{173}Yb	5/2	16.13	1.28	24.35	−1.30251	280
^{67}Zn	5/2	4.102	0.692	31.292	1.6766885	15
^{91}Zr	5/2	11.22	6.26	46.494	−2.49743	−17.6

Source: Data supplied by Bruker Spectrospin Ltd.

Note: Common nuclei are shown in bold; nuclei for which NMR spectra are unlikely to be observed are shown in italics. † Frequency to three decimals are experimental for International Union of Pure and Applied Chemistry (IUPAC) standards; frequency to two decimals are calculated from magnetic moments.

Intensities

When we irradiate the sample at the Larmor frequency we cause the nuclear spins both to be excited from lower to higher energy levels (absorption) and to relax from higher to lower states (emission). Since these absorptions and emissions are occurring with the same energy, i.e. at the same frequency, the intensity of the NMR signal depends on the difference between the number of absorptions and the number of emissions that occur as the resonance is excited. In NMR this depends, essentially, on the population difference between the upper and lower energy levels. We do not have to worry about extinction coefficients as we do in IR spectroscopy, for example; the probabilities of any given spin in the upper level falling to the lower and of any given spin being excited from the lower to the upper level are equal. We can therefore use the intensity of the NMR signal to count the number of nuclei giving rise to a transition, i.e. of a given type, in a molecule.

Unfortunately, the population difference in NMR is very small, a separation of energy levels of 10^{-25} J gives only a 1 in 50,000 population difference at 25°C at Boltzmann equilibrium (equation 1.4). NMR signals will therefore be very weak; NMR spectroscopy is a very insensitive technique. Nuclei with low gyromagnetic ratios will tend to give weaker signals than high γ nuclei, since the energy level separation depends on γ. Low sensitivity is NMR spectroscopy's greatest drawback. One way to improve the sensitivity is to make $\boldsymbol{B_0}$, the applied field, as large as possible. Using superconducting electromagnets fields up to 14.1 tesla are common in modern spectrometers with fields as high as 23.5 T available. To put this in context, a spanner 'lost' inside a 4.7 T magnet will support the weight of a (large) student!

NMR is insensitive because the nuclear magnetic moment interacts weakly with the environment. This makes the separation of the nuclear spin energy levels small and results in almost equal populations of the energy levels between which the NMR transitions occur (equation 1.4), hence NMR signals are very weak.

The separation of the NMR energy levels of nuclei with higher magnetogyric ratios (i.e. those resonating at higher frequencies) is larger (equation 1.2), so there will be a greater population difference resulting in stronger NMR signals for high γ nuclei.

$$\frac{n_{upper}}{n_{lower}} = e^{\frac{-\Delta E}{kT}}$$
$$\ldots\ldots \approx 1.00002$$

Eq. 1.4 The population difference between the upper and lower energy levels is given by the Boltzmann distribution and is very small.

Increasing the spectrometer magnet field strength also increases the separation of the energy levels (equation 1.2), increasing the population difference and resulting in a stronger NMR signal.

1.5 Nuclear spins and their surroundings

NMR linewidths and resolution

Nuclear spins interact only weakly with their surroundings *via* the magnetic fields produced by other magnetic dipoles such as electrons, paramagnetic species, other nuclear dipoles, and by nuclear quadrupoles. This means that the lifetimes of nuclear spin states (at least for spin ½ nuclei in solution) are long, relaxation is slow, and therefore NMR lines are very sharp—there is little Heisenberg broadening. For quadrupolar nuclei, however, the relatively strong interaction between the nuclear quadrupole and dipole shortens the lifetime of the excited state considerably, resulting in broader lines following Heisenberg's uncertainty principle (equation 1.5). Dipole–dipole interactions, chemical shift anisotropy, and other effects further broaden the lines in the solid state (see Chapter 7).

Paradoxically the weak interaction of the nuclear spin with its environment which gives both the small energy separation of the spin states and the long lifetimes is responsible both for the analytical power of the technique—separate resonances are observed for each chemically different site in a molecule, couplings between nuclei are well resolved, and so an accurate determination of the topology of a molecule is possible—and for the low sensitivity of NMR spectroscopy.

Eq. 1.5 For spin ½ nuclei, Δt is large: the uncertainty in the NMR resonance frequency, ΔE, is therefore very small since the lifetimes are long; NMR resonances (at least for spin ½ nuclei) are therefore very sharp.

$$\Delta E\, \Delta t = \hbar\,/\,2$$

Timescales

A compound must exist in a given energy state for a finite time in order to be detectable spectroscopically. This is sometimes referred to as the timescale of the spectroscopy. There are several ways to consider the timescale of a spectroscopy. For example, we might consider the effect of Heisenberg broadening on linewidths. If we take the energy separation we deduced using equation 1.2 above, 10^{-25} J, and substitute this into equation 1.5, we arrive at a required lifetime of around 10^{-9} s. This compares with an IR timescale, resonance frequency ~ 100 THz, of 10^{-13}–10^{-14} s. IR spectroscopy will be able to detect species having much shorter lifetimes than can NMR.

In NMR spectroscopy other useful timescales can be identified, for example, the ability to distinguish different, but exchanging species, say a gas adsorbing and desorbing on a surface, or a ligand in a complex exchanging with free ligand in solution. NMR experiments can study such processes over a wide range of timescales from seconds to fractions of milliseconds.

These timescales should not be confused with the experimental time needed to acquire an NMR spectrum, which can be anything from minutes to days depending on the experiment performed, the sensitivity of the nucleus under study, its natural abundance, and/or the concentration of the sample.

1.6 Chemical shielding

If the applied magnetic field, $\mathbf{B_0}$, were the only magnetic field experienced by the nuclear spin the NMR spectrum would be a rather uninteresting single line. Fortunately, the electrons in the molecule, since they are charged, produce additional magnetic fields as they move in the molecular orbitals, changing the field seen by the nucleus depending on its chemical environment. Two types of induced field occur. Circulation of electrons in the ground state orbitals produces a field that opposes the spectrometer magnet field (Figure 1.5), so the nucleus sees a smaller total field; we say the nucleus is *shielded* from the spectrometer field— this is called the diamagnetic term. Electrons mixed in from excited electronic states give rise to fields that can either reinforce or reduce the spectrometer field; the nucleus sees a greater or smaller field, respectively, and will be *deshielded* or *shielded*, respectively, from the spectrometer field. This contribution to shielding is called the paramagnetic term.

The diamagnetic and paramagnetic terms will depend, for example, on the elements and the type of bonding present, so will be characteristic of the chemical environment of the nucleus. The change in magnetic field experienced by the nucleus as a result of shielding changes the separation of the nuclear spin energy levels (equation 1.2), and hence the resonance frequency (equation 1.3); each chemically different group of atoms of a given NMR active element present will experience different shielding so see a slightly different magnetic field overall,

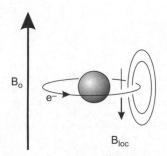

Figure 1.5 Electrons are charged and so experience a force causing them to move when placed in a magnetic field. This motion generates additional magnetic fields at the nucleus, giving rise to the chemical shift.

B_0

e^-

B_{loc}

i.e. will be shielded (or deshielded) differently. Put another way, the nuclear spin energy levels of each group of chemically equivalent nuclei will be split by a slightly different amount compared to every other group, and so will resonate at a slightly different frequency. This effect is **chemical shielding**.

Consider the hypothetical trigonal bipyramidal molecule AsF_3Me_2 (Figure 1.6). If we ignore any effects such as non-rigidity of the molecule, we can divide the fluorine atoms into two groups: one fluorine is in an equatorial position, F_e, and the other two are in the axial positions, F_a. Each group will have a different chemical shift due to the geometrically different positions they occupy. The ^{19}F NMR spectrum of AsF_3Me_2 will therefore show two sets of resonances, one for each type of fluorine. The relative total intensities of the sets of resonances reflects the number of fluorine atoms of each type; two $\times F_a$, one $\times F_e$. The resonance of each group of fluorine nuclei will be further split by scalar coupling, *vide infra*.

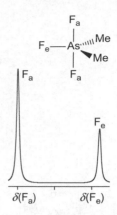

Figure 1.6 The ^{19}F NMR spectrum of AF_3Me_2 at low resolution shows separate resonances for the axial and equatorial fluorines.

The chemical shift and the ppm scale

Shielding arises from the interaction of the electrons surrounding the nucleus with the spectrometer field (Figure 1.5), so will change if the strength of the spectrometer field, $\boldsymbol{B_o}$, changes (equation 1.6), and so the resonance frequencies of all the nuclear spins in our sample will change (equations 1.2 and 1.3). Since different spectrometers use magnets of different field strengths, NMR resonance frequencies are normally reported as the fractional shift of the resonance line from that of a reference compound, the **chemical shift**. It is convenient to give this fractional shift in parts per million from a standard reference, the δ scale of chemical shift, since this gives the chemical shift in a form independent of the spectrometer used (equation 1.7). Recently, IUPAC has recommended that the protons of tetramethylsilane be used as a universal reference for all NMR spectra, regardless of the element/nuclide under study. To relate data on the unified scale to chemical shifts expressed relative to the traditional reference compound of the nuclide X, a quantity Ξ (Greek capital Xi) was defined as the ratio of the resonance frequency of the traditional reference, $\nu_{obs}(X)$, to that of ^1H of TMS in $CDCl_3$, $\nu_{obs}(TMS)$, in the *same* magnetic field (equation 1.7). This is both convenient and avoids the use of some nasty (extremely toxic) reference compounds, for example $HgMe_2$.

Shifts to higher energy, downfield, are positive, to lower energy, upfield, are negative. In old texts this convention may be reversed, particularly for nuclei other than proton and carbon. To convert from ppm to hertz simply multiply by the resonance frequency (at the spectrometer field) appropriate to the nucleus under study.

In inorganic chemistry it is difficult to predict the chemical shift, since many competing factors such as oxidation state, nature of the ligands, nature of the bonding, and charge, have to be taken into account. However, an empirical understanding of factors likely to affect the shift can be gleaned by study of data for

Eq. 1.6 The shielding constant σ is used to parameterize shielding. σ is dimensionless and characteristic of the chemical environment of the nucleus, i.e. σ is different for every different chemical environment but does not change when the spectrometer magnet is changed.

$$B = B_0(1-\sigma)$$

Eq. 1.7 The chemical shift is reported as the fractional frequency shift, in ppm, from the resonance frequency of a reference compound to remove the effect of the spectrometer magnetic field strength. Chemical shift increases as shielding decreases.

$$\delta = \frac{\nu_{observed} - \nu_{ref}}{\nu_{ref}} \times 10^6$$

Eq. 1.8 The unified scale of chemical shift. ν_{TMS} must be measured for every sample, requiring TMS to be added and a proton spectrum run at the same time as the spectrum of interest. Ξ can be looked up in tables.

$$\nu_{ref} = \Xi \times \frac{\nu_{TMS}^{measured}}{100}$$

known compounds. In this way the chemical shift can tell us about differences in oxidation state, the nature and position of bonded atoms, and about geometric factors—see Chapter 3.

NMR nomenclature

In predicting and analysing NMR spectra, it is usual to use the letters A, B, C, and so on to denote the different *groups* of spins present. Groups that are significantly different from A, B, C, etc are labelled X, Y, Z. This places the focus of attention correctly on the groups of spins present, rather than the individual spins or the elements present.

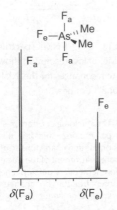

Figure 1.7 At high resolution the NMR resonances of the fluorines in AsF_3Me_2 are seen to be split by coupling between the two sets of inequivalent fluorines

Notice that we consider the two F_as as a group and work out the splitting caused by the group of spins; we do not consider the splitting caused by 'the first' F_a and then add 'the second' F_a noting that because the F_as are the same the coupling constant will be the same, so the two middle lines will overlap to give a 1:2:1 triplet.

As we shall in Chapter 2, when analysing and interpreting NMR spectra, thinking about individual spins, rather than about groups of spins, is unhelpful since it obscures the functional groups/structural elements present in the molecule.

The spins in a group do, of course, know that the other spins in the group are present, so coupling does occur. This can be thought of as the limiting case of an AB spectrum where the difference in chemical shifts has gone to zero; all the intensity ends up in a single inner line, and the outer signals having zero intensity, the coupling appears to vanish.

1.7 Scalar coupling

Whenever there are groups of different NMR active nuclei present in the molecule there will be scalar coupling between them. The nuclei can be of the same element—**homonuclear** coupling—or of different elements—**heteronuclear** coupling.

Scalar coupling arises because the spin of the observed nucleus senses the presence of the nuclear magnetic moments of other groups of NMR active nuclei surrounding it. These produce small 'extra' magnetic fields in addition to the fields due to the spectrometer magnet and the chemical shift. Thus, the observed nucleus sees not one resultant field but several, depending on the number and nature of the surrounding NMR active nuclei.

One important feature of scalar coupling is that it is seen *between groups* of different nuclei and not between nuclei within the same group. This means that identical nuclei—i.e. nuclei that occupy identical positions within a molecule, for example the two axial F_a nuclei in AF_3Me_2 (Figure 1.7)—do not show coupling to each other while coupling between the inequivalent axial and equatorial fluorines is seen (Figure 1.8). It is also important to remember that the splitting seen on a resonance is caused by the neighbouring groups, *not* by the group we are looking at, i.e. in Figure 1.7 the resonance of F_a is split by group F_e and the resonance of group F_e is split by group F_a.

It is important to realize that the mechanism of, and so the way to interpret scalar coupling is the same regardless of the nuclei involved. Thus, if we have a good physical picture of the way scalar coupling works and understand it we can deal with any coupling situation whether it be proton–proton, proton–carbon, silver–rhodium, boron–lithium, or whatever. We need, therefore, to focus on the groups of spins present and work out what splittings these will cause; we don't need to worry about the identity of the elements present.

The coupling constant

Scalar coupling depends only on the interaction of the nuclear magnetic dipoles, and not the spectrometer field. In contrast to chemical shielding, the coupling constant does not change if we go from one spectrometer to another and so is given in hertz not ppm. The separation of the resonance energies, in hertz, is called the (scalar) coupling constant, and is given the symbol J_{AB} or $J(AB)$. The chemical shift of nucleus **A** is calculated from the midpoint of the resonances in its coupled multiplet. Scalar coupling is reciprocal; the energy levels, and hence the NMR spectrum, of nucleus **B** will be split by the nuclear spin of **A**. The magnitude of the coupling from **A** to **B** will be the same as that from **B** to **A**, i.e. the coupling constant J_{BA} is the same as J_{AB} (Figure 1.8). Scalar coupling normally diminishes as the number of bonds between **A** and **B** increases, usually $^1J \gg {}^3J \geq {}^2J > {}^4J$. Beyond four bonds coupling is normally too small to be observed. The number of bonds between the coupling nuclei is indicated by a superscripted prefix; for example a coupling through two bonds is written $^2J_{AB}$. The magnitude of the coupling constant also varies with the angle between the bonds/coupling groups (see Chapter 3): *trans* couplings are usually larger than analogous *cis* couplings; J also varies with the dihedral angle between coupling groups reaching maxima at angles of zero and 180 degrees—the Karplus relation (equation 3.3).

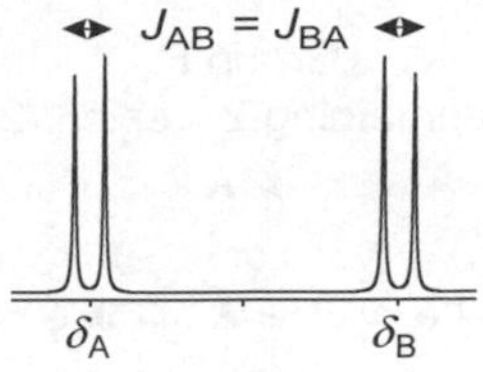

Figure 1.8 The coupling constant J_{AB} must equal J_{BA}. The chemical shifts δ_A and δ_B are measured from the midpoints of the multiplets.

Effect of changing the field on the appearance of the spectrum

There is an important difference in how shielding and scalar coupling respond when the spectrometer field is changed. Shielding results from the interaction of the electrons around the nucleus with the magnet field; this means that the shielding—so the resonance frequency of a multiplet—will change if we change the spectrometer. This is why we use the chemical shift scale, the chemical shift in ppm does not change when we change the spectrometer magnetic field. On the other hand, scalar coupling is not mediated by the spectrometer field but arises from the interaction of one nuclear spin with another—the size of the coupling, J, the coupling constant, does not change if we change the spectrometer. Working at very high fields separates out the multiplets, making it easier to distinguish resonances from closely similar groups (Figure 1.9).

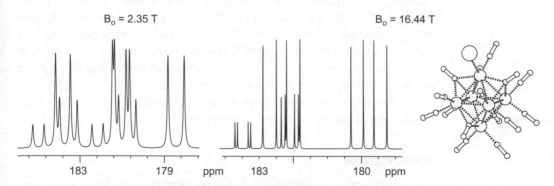

Figure 1.9 Working at high spectrometer field strengths 'stretches' the chemical shift axis but not scalar couplings, untangling overlapping resonances. Here, in the ^{13}C NMR spectrum of $Rh_6(CO)_{15}(PPh_3)$, the overlapping resonances of the terminal CO ligands are spread out by increasing the spectrometer field from 2.35 T to 16.44 T, greatly simplifying analysis of the spectrum, e.g. the doublet of doublets around 183.5 is clearly seen at the higher field. The spectrum recorded at 2.35 T is shown expanded seven times with respect to that recorded at 16.44 T to allow the separate resonances to be seen. The assignment of the spectrum is described in Chapter 3, Figure 3.17.

Group F_a
containing 2 x spin 1/2

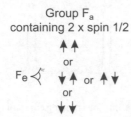

Figure 1.10 The spins in group F_e see the spins in the neighbouring group F_a either up or down. There are three distinguishable arrangements of the spins in group F_a so the resonance of F_e is split into a 1:2:1 triplet.

$P^AP h_3$
|
$Ph_3P^B – Rh — Cl$
|
$P^AP h_3$

$\delta(P^B)$

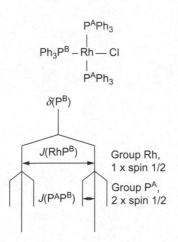

Figure 1.11 Stick diagrams can be used to represent the couplings in a spin system. Shown here is the stick diagram for the ^{31}P NMR resonance of the unique phosphine *trans* to chloride in Wilkinson's catalyst.

Coupling to a group of spin ½ nuclei

In our molecule AsF_3Me_2 we have two groups of fluorines, F_a and F_e. ^{19}F has spin ½ and group F_e has one fluorine in it; group F_e contains one spin ½. Group F_a, on the other hand, contains two fluorines—it is a group of two spin ½ nuclei.

In Figure 1.7 we can see that the resonance of group F_a is split by the spin ½ in group F_e into a doublet. This is because F_a can see the spin ½ in group F_e either aligned with the magnet field or against the magnet field. These two alignments have different energies, so the resonance of F_a is split in two. Why then does the resonance of F_e split into a 1:2:1 triplet? We need to look at the possible arrangements of the two spin ½ nuclei in group F_a. There are three possible arrangements (Figure 1.10)—both spins up, both spins down, or one spin up and one spin down. Since the two F_a spins are indistinguishable, the two possible ways to get one spin up and one spin down are also indistinguishable. The chance of having one up and one down is thus twice that of having both spins up and also twice that of having both down—so our three arrangements occur in the ratio 1:2:1. A group of two spin ½ nuclei splits its neighbour into a 1:2:1 triplet. Notice that we deduced the splitting pattern by considering the *spins* in the neighbouring group. The fact that these belong to ^{19}F is not needed; *any* group of two spin ½ nuclei will split its neighbour into a 1:2:1 triplet.

For coupling to a group of several spin ½ nuclei the result above may also be obtained using Pascal's triangle.

Coupling to groups of inequivalent nuclei

If there are several neighbouring groups of spins, we can work out the coupling patterns these cause simply by considering each group of equivalent spins in turn. Of course, we need to remember that the size of the splittings caused by the different groups of spins (the coupling constant *J*) will not be the same. Consider, for example, the ^{31}P NMR spectrum of Wilkinson's catalyst $[Rh(PPh_3)_3Cl]$. We can see (Figure 1.11) that there are two types of ^{31}P present, a group of two mutually *trans* Ps (a group of two spin ½ nuclei, so will split any neighbouring group into a 1:2:1 triplet) and a group of one P *trans* to Cl (a group containing one spin ½ nucleus, so will cause doublet splittings). There is also a group of one ^{103}Rh (another group containing one spin ½ nucleus that will cause doublet splittings in its neighbours).

To keep track of all the couplings, we use stick diagrams in which each layer details the splitting caused by one of the *groups* of spins.

Stick diagrams

Stick diagrams are a convenient way to keep track of the couplings on a given resonance. Let's look at the splittings expected on the resonance of P^B in Figure 1.11 caused by the neighbouring groups Rh and P^A. Starting from a single line to indicate the chemical shift of P^B we consider the splitting caused by each group of coupling nuclei in turn, writing in new lines to represent the number and intensity of the resultant spectral lines. We know that a group of one spin ½ nucleus

(Rh) will split its neighbour into a doublet. The group of two spin ½ nuclei (P^A) then splits each of the resultant lines into a 1:2:1 triplet (Figure 1.11). The resonance of P^B will be a doublet of triplets. It does not matter in which order we consider the couplings, nor does it matter how big we draw the splittings, although it is usually simpler to start with the biggest and work down.

The stick diagram is only intended to keep track of the splittings: each group of neighbouring spins is represented by a layer in the diagram; the diagram is a topological representation of the spin system and is not supposed to 'look like' the multiplet in the experimental spectrum, just as the map of the London Underground is not supposed to look like the layout of the streets London.

Decoupling

The effect of scalar coupling can be removed by irradiating the coupling nucleus, say nucleus **A**, at its resonance frequency to induce rapid transitions between the possible orientations of the spin on **A**. Neighbouring nuclei then see a single, averaged field, rather than distinct fields, for the different possible orientations of the nuclear magnetic moment of **A**. The coupling appears to vanish. By convention we write the decoupled nucleus in braces {curly brackets}; thus, a proton-decoupled boron-11 spectrum is written ^{11}B{^{1}H}.

It is also possible to decouple selectively one set of, say fluorine, nuclei whilst leaving others unaffected. For example, in our AsF$_3$Me$_2$ example above ('Coupling to a group of spin nucleus') we could selectively decouple the axial fluorines; the triplet due to the equatorial fluorine then collapses to a singlet (Figure 1.12). In this way it is possible to simplify apparently complex spectra and identify exactly which nuclei are responsible for any given coupling in the spectrum.

It can be helpful in interpreting NMR spectra to remove the coupling due to one or more neighbouring groups of spins.

Decoupling experiments are useful not only to simplify the spectra (e.g. ^{13}C and ^{31}P NMR spectra are routinely acquired with proton decoupling), but are also used to confirm which neighbouring group causes which coupling.

In stick diagram terms, decoupling removes one (or more) layer(s) from the stick diagram.

1.8 Quadrupolar nuclei, spins > ½

We can easily extend our physical picture from spin ½ nuclei to those with spins greater than ½. A nucleus of spin **I** has $2I + 1$ m_I levels associated with it. When placed in a magnetic field, each m_I corresponds to a different orientation of the

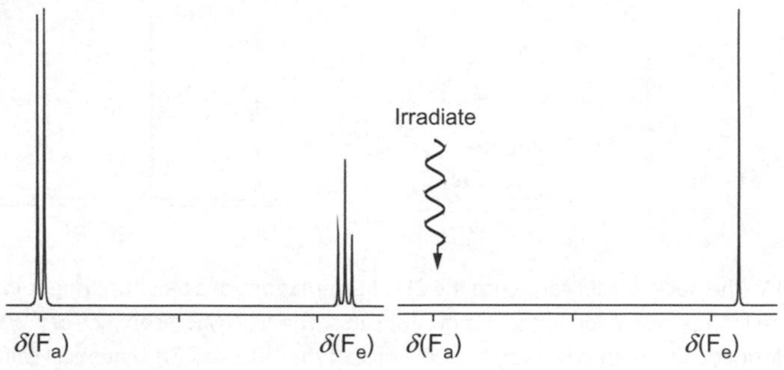

Figure 1.12 Irradiation of the axial fluorines removes the coupling from F$_a$ to all other nuclei in the molecule

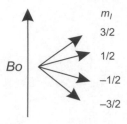

Figure 1.13 Placing a quadrupolar nucleus in a magnetic field, B_o, lifts the degeneracy of the m_I states giving $2I + 1$ different energy levels, each corresponding to one of the allowed orientations of the nuclear spin with respect to the applied field. Here the four m_I states for a spin $I = \frac{3}{2}$ nucleus are shown.

Figure 1.14 The energy levels are equally spaced. The selection rule $\Delta m_I = \pm 1$ results in all allowed transitions having the same energy.

nuclear spin to the field. For example, a spin $\frac{3}{2}$ nucleus such as ^{11}B will give four energy levels when placed in a magnetic field B_o corresponding to the four allowed m_I values; there are four allowed orientations of the nuclear magnetic moment with respect to the applied field (Figure 1.13). Inspection of equation 1.2 tells us that these energy levels will be equally spaced (Figure 1.14). Since the NMR selection rule is $\Delta m_I = \pm 1$ we immediately see that all allowed transitions will have the same energy—a single line at the chemical shift will be observed for an isolated quadrupolar nucleus—at least in solution (see Chapter 7 for solid-state effects).

Coupling to nuclei with spin >½

We saw in Figure 1.10 that a nuclear spin (let's call it **A**) can sense the orientations of a neighbouring spin (let's call the neighbour **B**) and that this causes splitting of the NMR resonance of **A** by **B**. The number of allowed orientations of a nuclear spin I is $2I + 1$, so a neighbour of spin I will split the resonance of **A** into $2I + 1$ equally intense lines, since each possible m_I state of **B** occurs with almost equal probability. Suppose **B** has spin $\frac{3}{2}$. $2I + 1$ is then four, m_I of **B** can take the values $+\frac{3}{2}, +\frac{1}{2}, -\frac{1}{2}$ or $-\frac{3}{2}$ (Figure 1.13). **A** will sense these four possible orientations of the **B** spin and so its resonance splits into four equally intense lines.

An example of such a spin system is the ^{1}H NMR spectrum of $[^{11}BH_4]^-$, (**A** = ^{1}H; **B** = ^{11}B). Note that in $[^{11}BH_4]^-$, all four protons are in identical environments (i.e. in the same group), so coupling between them is not seen; we only need consider coupling to the ^{11}B. The ^{1}H NMR spectrum of $[^{11}BH_4]^-$ shows four equally intense resonance lines due to coupling to the spin $\frac{3}{2}$ ^{11}B. The chemical shift is the midpoint of the multiplet (Figure 1.15).

Coupling to several identical quadrupolar nuclei

When coupling occurs to a group containing several quadrupolar nuclei (i.e. several, equivalent nuclei with spins >½), we just need to consider the combinations

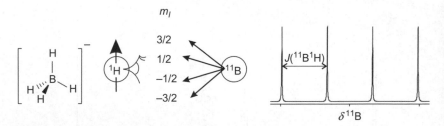

Figure 1.15 The nuclear spin can sense the $2I + 1$ orientations of a neighbouring spin. This is shown here for a proton coupling to a spin $\frac{3}{2}$ nucleus. The ^{1}H NMR resonance of $[^{11}BH_4]^-$ is split into four equally intense lines by the presence of the ^{11}B, $I = \frac{3}{2}$. The chemical shift of the protons is the mid-point of the multiplet.

of spin orientations, m_I values, possible and work out the number of distinguishable permutations possible. We use the same method as we did for two spin ½ nuclei in Figure 1.10.

Consider the bridging hydrogens in $Me_4{}^{11}B_2H_2$; the spin system is A_2B_2, $A = {}^1H$, $B = {}^{11}B$; for simplicity's sake let's assume the methyl groups are too far away to couple. The bridging protons, spin ½, are in identical environments—so coupling between them is not seen. Similarly coupling is not seen between the identical, spin ³⁄₂, borons. There will, however, be coupling between the borons and protons. The boron resonance will be split into a 1:2:1 triplet by the group of two equivalent proton spins following Pascal's triangle rule. To work out the splitting pattern of the proton resonance caused by the group of two equivalent ^{11}B spins, we work out all possible combinations of the boron spin orientations just as we did for the proton spins in Figure 1.10. Since the borons are equivalent, combinations that give the same Σm_I will be indistinguishable. The proton resonance will thus split into a 1:2:3:4:3:2:1 septet according to the different combinations of the boron spins, as shown in Figure 1.16.

Coupling to several groups of neighbours, including quadrupolar nuclei

If several neighbouring groups of spins are present, including groups of quadrupolar nuclei, we can use the same method as we did for coupling to several groups of spin ½ nuclei: we just deal with each group in turn and use a stick diagram to keep track of the splittings.

Consider $Me_2{}^{11}B(\mu–H)_2{}^{11}BH_2$ (Figure 1.17). There are four chemically different groups of spins in the molecule; the terminal hydrogens (a group of two spin ½ nuclei), the bridging hydrogens (another group of two spin ½ nuclei), and the two different boron atoms (two groups, each containing one spin ³⁄₂ nucleus); again, let's ignore the methyl groups to keep things simple. Now let's work out what we expect the ^1H NMR resonance of the terminal hydrogens, H^t, to look like and use a stick diagram to keep track of the couplings. H^t will be split by the directly bonded boron $^{11}B^A$ (spin ³⁄₂) into four equally intense lines, $^1J_{BH}$. Each line will then be split into four more equally intense lines by the remote boron-11 spin, $^{11}B^B$, $^3J_{BH}$. Each of the resultant lines will be further split into 1:2:1 triplets by the bridging hydrogens, H^b (a group of two equivalent spin ½ nuclei, Pascal's triangle rule), $^2J_{HH}$. The NMR resonance of the terminal hydrogens is thus split into forty-eight lines of intensities $(1:2:1) \times 4 \times 4$, Figure 1.17.

Quadrupolar broadening

Not all quadrupolar nuclei give useful NMR spectra. This is because the nuclear quadrupole can cause very rapid relaxation of the nuclear spin states. Short lifetimes result in extreme Heisenberg broadening and hence very broad—many kilohertz—linewidths. For similar reasons, coupling to quadrupolar nuclei is often not seen. The efficiency of quadrupolar relaxation, and hence the degree

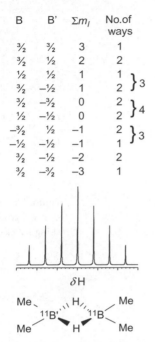

B	B'	Σm_I	No. of ways
³⁄₂	³⁄₂	3	1
³⁄₂	½	2	2
½	½	1	1
³⁄₂	–½	1	2 } 3
³⁄₂	–³⁄₂	0	2
½	–½	0	2 } 4
–³⁄₂	½	–1	2
–½	–½	–1	1 } 3
³⁄₂	–½	–2	2
³⁄₂	–³⁄₂	–3	1

δH

Figure 1.16 To determine the splitting pattern of a nucleus coupled to a group of more than one equivalent nucleus, we simply work out the number of different combinations of the coupling spins possible. In the case of equivalent spins, the splitting caused by any given spin is the same as that produced by any other, so combinations that give the same value for Σm_I are indistinguishable. Two equivalent boron spin ³⁄₂ neighbours will split the proton resonance into a 1:2:3:4:3:2:1 septet—see text.

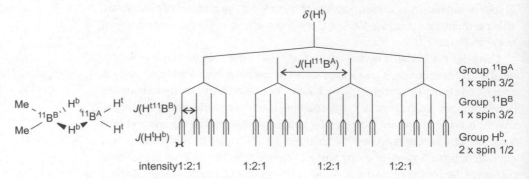

Figure 1.17 Predicting the splitting pattern for spin systems containing several groups, involving both spin ½ and spin > ½ groups of nuclei is straightforward; simply consider each group in turn and keep track of the splittings using a stick diagram

of broadening observed, depends on several factors, including: (1) the symmetry of, or more strictly, the electric field gradient (EFG) at, the site—quadrupole broadening is much reduced in cubic symmetry; and (2) the rate of molecular tumbling—rapid tumbling helps average out the effect of the EFG. For this reason, the NMR spectra of quadrupolar nuclei in solution are often measured at elevated temperature. This is illustrated in the spectra of some zirconocene complexes (Table 1.3), in which the increase in linewidth closely follows the increase in EFG as the substituent on Zr goes from Br to Me, whereas, for chloride complexes, where changes in the EFG are less pronounced, the linewidth increases

Table 1.3 Variation in ^{91}Zr linewidth with molecular symmetry and tumbling rate

Complex	Half-height line width/Hz	Relative EFG	Relative molecular tumbling rate
300 K	–	–	–
Cp_2ZrBr_2	16	0.5	–
Cp_2ZrCl_2	270	1.0	1.0
Cp_2ZrMe_2	2530	9.3	–
$(C_5H_4Me)_2ZrCl_2$	540	0.7	1.8
$((t-Bu)C_5H_4)_2ZrCl_2$	2810	2.0	4.9
$(C_5H_4Me)_2ZrMe_2$	4030	–	–
$((t-Bu)C_5H_4)_2ZrMe_2$	8250	–	–
348 K	–	–	–
Cp_2ZrMe_2	1070	–	–
$(C_5H_4Me)_2ZrMe_2$	1630	–	–
$((t-Bu)C_5H_4)_2ZrMe_2$	3230	–	–

Source: From data in Buhl et al. (1996), *Organometallics*, **15**, 778–85.

as the bulkiness of the cyclopentadienyl ring substituent increases, so tumbling rate decreases.

Remember, though, that not all quadrupolar nuclei suffer from extreme broadening; we saw above examples of the use of ^{11}B NMR to characterize diborane complexes; indeed ^{11}B NMR is routinely used to characterize quite complex boron cluster compounds, since well-resolved resonances are found for each type of boron in the molecule and couplings to protons are usually well resolved. Other quadrupolar nuclei which give useful NMR spectra include ^{2}D, ^{6}Li, ^{17}O, ^{23}Na, ^{27}Al, ^{51}V, ^{59}Co, and ^{133}Cs, all of which have quadrupole moments less than $\pm 0.15 \times 10^{-28}$m^2.

1.9 Natural abundance

Some elements possess more than one isotope, not all of which will be NMR active. This situation is familiar from organic chemistry; carbon has three isotopes: ^{12}C, which is NMR inactive and makes up 98.9% of natural carbon; ^{13}C, which has spin ½ and is 1.1% abundant; and ^{14}C, which is NMR inactive. The NMR active isotopes of an element may have different nuclear spins, I, and will certainly have different gyromagnetic ratios, γ, and hence different resonance frequencies (Table 1.2). The different isotopes, therefore, will normally appear in their own spectrum and will show different coupling constants and cause different coupling patterns on their neighbours, since these change with γ and I. Since only some of the sample will contain the isotope of interest, the NMR signal is reduced in strength, which can be problematic for signal to noise. In some circumstances, it is worthwhile using isotopically enriched samples to boost the NMR signal; for example, it is quite common to enrich the carbonyl ligands of transition metal carbonyl compounds with ^{13}CO and to use isotopic enrichment when studying ^{17}O or ^{15}N.

Satellites

The presence of a coupling isotope that is less than 100% abundant affects the appearance of the NMR spectra of other nuclei in the sample. The simplest way to think of this is to divide, in our mind, the sample into those molecules that contain the <100% abundant NMR active isotope and, in a separate pot, the molecules that do not. The molecules in the first pot will show coupling to this nuclide, while those in the latter pot will not. To get the overall NMR spectrum, we add together the spectra of the two pots, weighted according to the natural abundance. The result is a central, unsplit resonance from the molecules that do not contain the NMR active coupling isotope and **satellites** on either side of this central resonance resulting from coupling to the <100% abundant nuclide.

For example, ^{29}Si, has a spin ½ and is 4.7% abundant. The ^{19}F NMR spectrum of SiF$_4$ is shown in Figure 1.18. A strong central peak is seen due to molecules that contain NMR inactive silicon. A doublet due to those molecules that contain ^{29}Si can be seen on either side of the main resonance, the ^{29}Si satellites.

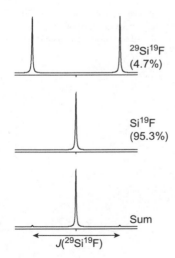

Figure 1.18 The ^{19}F NMR spectrum of SiF$_4$ shows satellites due to coupling to ^{29}Si; top–molecules containing ^{29}Si only, middle–molecules containing NMR silent ^{28}Si and ^{30}Si, and bottom–weighted sum of both.

Isotopologues

Finally, we need to consider what happens when an element possesses more than one NMR active isotope. In the examples above, the boron was quite intentionally specified as the isotope ^{11}B. This was because when there is more than one atom of the isotopically diverse element present, we must take account of the different possible combinations of isotopes that may arise—the **isotopologues**—and the abundance of each. This adds further complexity to the NMR spectrum (just as it does in mass spectrometry) but this can easily be dealt with by the approach outlined in 'Satellites' immediately above. If our molecule contains isotopically diverse nuclei we separate, mentally, the isotopologues into separate pots and work out the NMR spectrum for each isotopologue separately, then add these together taking account of the amount of each isotopologue expected. Fortunately, the chemical shift of the observed nucleus is fairly insensitive to isotope effects from adjacent, coupling nuclei so the centres of the multiplets do not vary greatly between isotopologues.

For example, the ^1H NMR spectrum of the BH_4^- ion at natural abundance is worked out in Figure 1.19. First consider the 80% of molecules that contain ^{11}B; the ^1H NMR spectrum will be four lines of equal intensity due to coupling to the spin ½ ^{11}B (Figure 1.19a). Similarly, for the 20% of molecules that contain ^{10}B the ^1H NMR spectrum will be seven lines of equal intensity due to coupling to the spin 3 ^{10}B ($I = 3$, $2I+1 = 7$ possible orientations) (Figure 1.19b). Finally, we must add the spectra of the ^{11}B and ^{10}B containing molecules in the ratio 80:20 (following the natural abundances of the two isotopes) to obtain the ^1H NMR spectrum of BH_4^- (Figure 1.19c).

Notice that the coupling constants to ^{11}B and ^{10}B are different. In fact, the ratio of the coupling constants is the ratio of the gyromagnetic ratios of the two isotopes (equation 1.9).

If there are several isotopically diverse nuclei, for example the diboranes discussed above, we must take account of the different possible combinations

---

Isotopologues are molecules that have the same chemical formula and bonding arrangement of atoms but differ in their isotopic composition. Isotopomers have the same chemical formula and bonding arrangement of atoms and isotopic composition but differ in where different isotopes of the same element are located in the molecule. In common parlance, 'isotopomer' is used incorrectly when 'isotopologue' is meant.

In Figure 1.19 we notice that $J(^{10}BH)$ is much smaller than $J(^{11}BH)$. This is because the magnitude of the coupling constant depends on the magnetogyric ratio. If only the isotope of the coupling nuclide, so only γ has changed, there is a direct relation between the ratio of the γs and the Js:

$$\frac{\gamma(^{11}B)}{\gamma(^{10}B)} = \frac{\gamma(^{11}BH)}{\gamma(^{10}BH)}$$

Eq. 1.9 We can calculate the ratio of the magnetogyric ratios directly from the ratio of the coupling constants and vice versa.

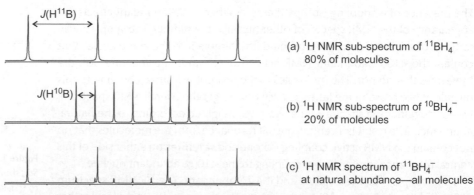

(a) ^1H NMR sub-spectrum of $^{11}BH_4^-$ 80% of molecules

(b) ^1H NMR sub-spectrum of $^{10}BH_4^-$ 20% of molecules

(c) ^1H NMR spectrum of $^{11}BH_4^-$ at natural abundance—all molecules

Figure 1.19 The NMR spectrum of a molecule in which coupling to an element possessing several NMR active isotopes occurs is calculated by considering coupling to each isotope separately then adding the sub-spectra in the ratios of the natural abundances of the isotopes.

Table 1.4 The possible isotopologues of a compound containing two boron atoms

Isotopologue	Probability		
$^{11}B^{11}B$	0.8×0.8	=	0.64
$^{11}B^{10}B$	$2 \times 0.8 \times 0.2$	=	0.32
$^{10}B^{10}B$	0.2×0.2	=	0.04

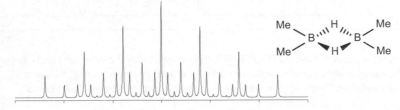

Figure 1.20 The ^1H NMR spectrum (B_2H_2 region) of $B_2H_2Me_4$ at natural abundance shows coupling due to all the boron isotopologues. The major resonance, a 1:2:3:4:3:2:1 septet is due to the $^{11}B^{11}B$ isotopologue. An overlapping 1:1:1:1 quartet of 1:1:1:1:1:1:1 septets of intermediate intensity, is due to the $^{11}B^{10}B$ isotopologue, whilst the very weak peaks are due to the $^{10}B^{10}B$ isotopologue.

Source: From data in R. E. Williams et al. (1960), *J. Phys. Chem.* **64**, 1583.

of boron isotopes, ^{10}B (20% abundant) and ^{11}B (80% abundant) (Table 1.4), and work out the NMR spectrum expected for each isotopologue.

Thus, the observed NMR spectrum for the bridging protons in $Me_2B(\mu-H)_2BMe_2$ will be a superposition of the spectra resulting from the isotopologues $Me_2{}^{11}B(\mu-H)_2{}^{11}BMe_2$, $Me_2{}^{11}B(\mu-H)_2{}^{10}BMe_2$, and $Me_2{}^{10}B(\mu-H)_2{}^{10}BMe_2$, weighted according to the probability of each occurring (Figure 1.20). Although this spectrum looks horrendously complicated, we can understand how it arises using our simple picture of couplings outlined above and Table 1.4.

1.10 Second-order couplings

So far we have considered only *first-order* couplings, where the splitting pattern can be deduced in a straightforward fashion as described above and the line separations relate directly to the coupling constants. However, some spin systems do not obey these simple rules and it is necessary to consider *second-order* effects. Analysis of second-order spin systems to obtain values for the chemical shifts and coupling constants is beyond the scope of this book; spectral simulation programs are available to do this.

In practice it is essential that we recognize when second-order effects are occurring, since in a **second-order spectrum** it is probable that none of the line separations corresponds to a coupling constant; rather, the lines are split by the sums and differences of the coupling constants. Second-order spectra

Second-order effects are commonly encountered when coupled groups of spins have similar chemical shifts (roofing) or when the spins in one group of chemically equivalent spins couple differently to spins in another group (magnetic inequivalence).

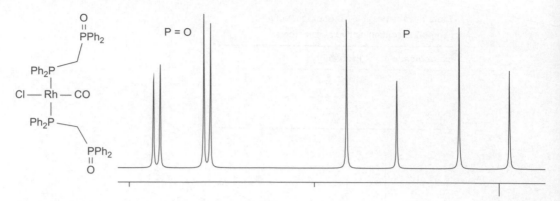

Figure 1.21 The $^{31}P\{^{1}H\}$ NMR spectrum of *trans*-Rh(CO)Cl{Ph$_2$PCH$_2$P(O)Ph$_2$}$_2$ is distorted by second-order effects due to the similar chemical shifts of the two mutually coupled phosphorus nuclei. This spectrum shows 'roofing', the tendency of the inner lines in a coupled system to 'steal' intensity from the outer lines. Roofing is an indication of the onset of second-order effects.

commonly occur when two coupled have very similar chemical shifts, for example the familiar **AB** quartet encountered for diastereotopic CH$_2$ groups. Second-order effects initially show up as roofing of the peaks in the spectrum; the inner lines 'steal' intensity from the outer lines, as in the ^{31}P NMR spectrum of the rhodium complex shown in Figure 1.21, in which the phosphorus nuclei of the co-ordinated phosphine and of the phosphine oxide ends of the ligand are coupled and both have shifts close to 30 ppm.

Chemical vs magnetic equivalence

Second-order spectra are also observed when chemically equivalent nuclei are magnetically inequivalent—this can happen in apparently very simple molecules. The concept of chemical equivalence is straightforward: groups should be chemically the same. **Magnetic inequivalence** occurs when the spins in a group of chemically equivalent nuclei couple differently to the spins in a neighbouring group. For example, the phosphorus atoms in [Pd(dipp)F$_2$] are chemically equivalent, as are the fluorine atoms; however, P^A is *trans* to one fluorine, F$^{A'}$, but *cis* to the other, F^A. Clearly P^A must couple differently to the two chemically equivalent fluorines. This is magnetic inequivalence and results in additional splittings in the NMR spectrum (Figure 1.22).

Magnetic inequivalence can result in extremely complicated NMR spectra for apparently simple molecules; for example, the $^{31}P\{^{1}H\}$ NMR spectrum of Mo(CO)$_3$(P$_6$) shown in Figure 1.23.

Although the complexity of the spectrum should dissuade all but the most fanatical from trying to determine the coupling constants, enough information can be obtained from a cursory inspection of the spectrum to establish the likely conformation of the P$_6$ ligand. Clearly there are three types of phosphorus, with two types being similar and one type very different. We can therefore deduce that the P$_6$ ring ligand must be in a boat conformation with four co-ordinated and two unco-ordinated phosphorus atoms rather than in a crown conformation

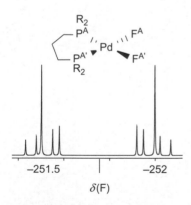

Figure 1.22 Second-order spectra can be observed in apparently simple molecules. *Magnetic inequivalence* occurs when the spins in a group of chemically equivalent nuclei couple differently to the spins in a neighbouring group.

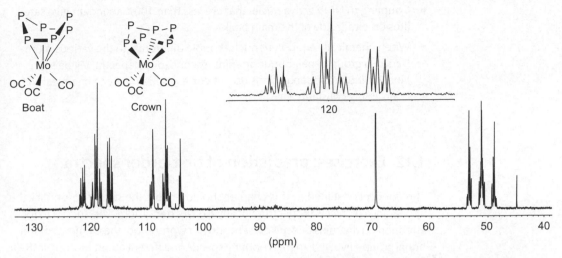

Figure 1.23 Second-order effects result in a very complicated $^{31}P\{^1H\}$ NMR spectrum (inset is an expansion of the resonance centred around 121 ppm) for the apparently simple $Mo(CO)_3(P_6)$ molecule due to extensive couplings between the magnetically inequivalent, but chemically equivalent, phosphorus atoms. Nevertheless, the structure of the compound can be established without fully interpreting the spectrum: there are three pairs of chemically equivalent, but magnetically inequivalent phosphorus centres implying a twisted boat, rather than a crown conformation for the P_6 ligand.

Source: Data provided by Dr M. J. Mays, University of Cambridge, personal communication. There is an impurity at *c.* 70 ppm.

with three co-ordinated Ps. There is an important point here; it is not necessary to understand every last coupling in an NMR spectrum to obtain the desired structural information; indeed, sometimes it is pointless to search for an understanding of every coupling.

1.11 Summary

- Understanding NMR spectroscopy is about understanding the behaviour of the nuclear spins present, not about the elements present.

- Understanding how spins behave allows us to understand the NMR spectrum of any (NMR active) element and to *predict* the appearance of first order NMR spectra.

- We should think in terms of the groups of equivalent spins present, not about individual spins.

- Each group of spins will give a resonance at a chemical shift that reflects the chemical nature of its surroundings.

- *J* coupling is due to the nuclear spins at the site under observation seeing the different arrangements of the spins of neighbouring groups.

- Quadrupolar nuclei often give very broad NMR lines, too broad to be easily detected, due to efficient quadrupolar relaxation. Scalar couplings to quadupolar nuclei are often not seen.

- Coupling to NMR active nuclei that are less than 100% abundant give satellites on either side of the main peak.
- When chemically equivalent nuclei couple differently to the various members of a group of neighbouring spins, second-order spectra will be seen. Line positions and separations do not correspond directly to chemical shifts or coupling constants.

1.12 Exercises: prediction of first-order spectra

The concepts outlined in this chapter should enable the reader to predict the appearance of the NMR spectrum of any **first-order** system. The 'trick' is to focus attention on the spins that are present and not worry about the elements or functional groups involved; there is nothing special or different about proton NMR—it is just another spin ½ nucleus subject to the same 'rules' as any other spin ½. Perhaps the most important point to remember in dealing with the NMR spectra of other spin ½ nuclei is the converse of this, the rules governing the appearance of the NMR spectrum are the same as those that govern ^{1}H NMR spectra, so if we understand those rules and where they come from, we understand the NMR spectra of all spin ½ nuclei and can extend our understanding to spins > ½.

Workflow

- Identify the NMR active nuclei.
- Divide these into groups based on chemical equivalence.
- Everything that follows will be based on these groups of spins.
- Predict the number of chemical shifts expected in each spectrum.
- Identify the neighbouring groups and predict the splitting each group will cause based on the spins contained in the neighbouring group.
- Construct a stick diagram to show the result.

1. In this exercise we will predict the appearance of some NMR spectra knowing only the spins present. A, B, and C have spin ½. X has spin 1 and Y spin ³⁄₂. Assume that all groups of spins couple to all other groups of spins present.

    a. Draw stick diagrams to represent the coupling patterns at A in the following spin systems:AC_2, AB_3, AB_2X, AB_2XY.

    b. What will the coupling pattern at X be in AB_2X? At Y in AB_2XY?

2. For each of the compounds below:

    a. Identify ALL NMR active nuclei.

    b. If a nucleus is not 100% abundant, indicate the natural abundance and show the satellites, except for ^{13}C, in your answers.

    c. Where quadrupolar nuclei occur, indicate their presence.

N.B. In practice the resonances of, and couplings to, quadrupolar nuclei are often not seen. You should ignore quadrupolar nuclei in the rest of this exercise unless you are explicitly asked to comment on them.

    d. Group the nuclei you have identified according to the NMR equivalence and hence decide on the number of chemical shifts that will be found in each nucleus's NMR spectrum.

Hint: Look at the *trans* group when deciding whether nuclei are equivalent or not.

    e. Draw stick diagrams for the elements specified to indicate where coupling will occur and the splittings that result.

Hint: Each level/layer in the stick diagram should represent the splitting caused by a neighbouring *group* of spins, NOT the individual spins within a neighbouring group.

Hint: For additional practice, try constructing stick diagrams for the other nuclei.

N.B. The NMR spectra of phenyl rings in phosphine ligands are usually complex. You should ignore the phenyl rings in this exercise.

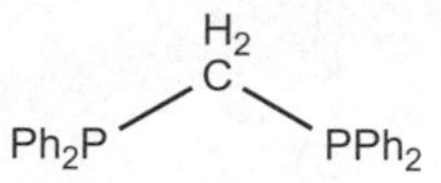

bis-diphenylphosphinomethane
stick diagram: phosphorus

trifluoromethylselenomercuric chloride
stick diagram: mercury

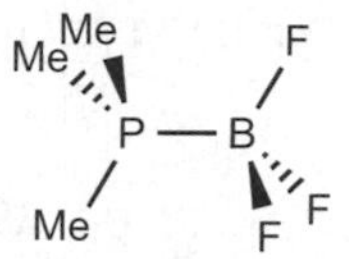

Me$_3$P.BF$_3$
stick diagrams: fluorine and boron
Exclude Me groups, include boron

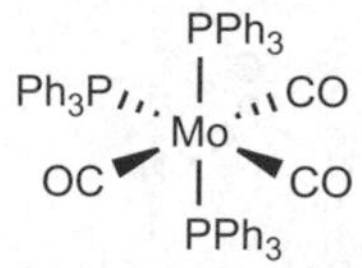

mer-tris-triphenylphosphine-
molybdenumtricarbonyl
stick diagram: phosphorus

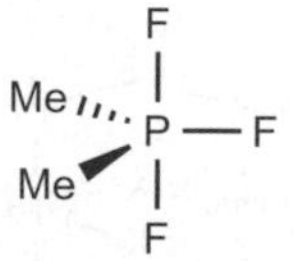

dimethylphosphorustrifluoride
stick diagram: phosphorus
Exclude the methyl groups

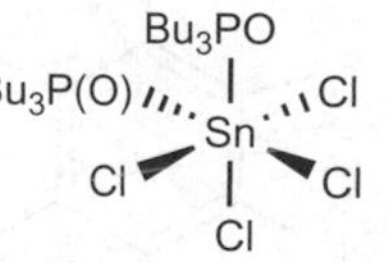

cis-bis-tributylphosphineoxidetin(IV)Cl$_4$
stick diagrams: tin and phosphorus
Exclude the butyl groups

3. Which of the following stick diagram(s) accurately represents the groups of spins that couple to the central carbon, C^B, in $CF_3C^BF_2CF_2OClO_3$? Coupling to ^{13}C at natural abundance is not shown.

4. Identify which of the following molecules will show second-order NMR spectra.

5. Write down stick diagrams that represent the couplings present in the ^1H NMR spectrum of $[Cp_2W(H)(SnCl(t\text{–}Bu)_2)]$, hydride region only, shown below, Figure 1.24. Exclude couplings to the Cp rings. The truncated central peak is a singlet and accounts for approximately 70% of the total intensity.

Hint: First construct separate stick diagrams for each isotopologue.

3.

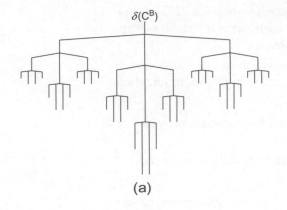

(a)

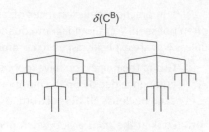

(b)

δ(C^B)

(c)

δ(C^B)

(d)

4.

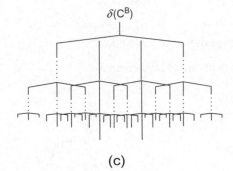

1,2,4,5-tetrafluorobenzene

X = acetylide

Hexachlorotriphosphazene

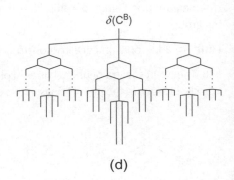

fac-tris-Triphenylphosphinemolybdenumtri(^{13}CO)

Diborane

trans or *gauche*-Tetrafluorohydrazine

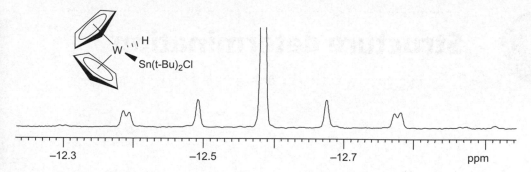

Figure 1.24 The ^{1}H NMR spectrum of [$Cp_2W(H)(SnCl(t\text{–}Bu)_2)$], hydride region only.

Source: Adapted from Mobley, T. A., et al. (2010), *Magn. Reson. Chem.*, **48**, 787–92. doi:10.1002/mrc.2663. Copyright © 2010 John Wiley & Sons.

2 Structure determination

2.1 Introduction

Chapter 1 presented some essential theory of NMR in solution together with a formalism to *predict* the NMR spectrum of a compound/spin system. Emphasis was placed on the fact that the appearance of the spectrum depended on the groups of nuclear spins present and not on the particular elements present.

In this chapter we explore the converse of prediction, the *interpretation* of NMR spectra to deduce the structure of the compound. This of course is the problem that the practising chemist will usually encounter; the hope is that the understanding of the basic principles gained in Chapter 1 will inform our chemist's interpretation of the spectra we meet in this. We will find that the same interpretation rules apply whether we are studying an 'organic' or 'inorganic' compound and whether we are looking at a proton NMR spectrum or that of any other nuclide; it is only the chemistry that changes, not the NMR!

2.2 Interpretation of an NMR spectrum

Workflow

- Look for symmetrical patterns and so identify the number of chemical shifts—this tells us the number of chemically different sites/groups of spins of the nuclide observed.
- Identify the coupling pattern/multiplicity of the resonance of each group/chemical shift.
- Identify the neighbouring group responsible for each coupling pattern using the coupling constant.
- Use the multiplicity produced by the neighbouring group in the resonance under observation to determine the number and type of spins in the neighbouring group.

Hints

- Don't guess a structure then try to force fit the NMR spectrum/spectra to it.
- Consider the groups of spins present, not individual spins.
- It is important to remember that coupling is caused by neighbouring groups of spins, and the patterns produced tell us about the membership of the neighbouring group, not about the membership of the group whose resonance we are looking at.
- We may have NMR spectra of more than one nuclide to consider; coupling may be homonuclear—the coupling partner is of the same nuclide, so its resonance appears in the same spectrum—or heteronuclear—the coupling partner is of a different nuclide, so its resonance appears in 'the other' NMR spectrum. The rules for interpreting homo- and heteronuclear couplings are the same.

Caveats

- Only first-order patterns are truly symmetrical.
- Second-order effects can distort the intensities of the lines within a multiplet so that they are no longer perfectly symmetrical. This is called roofing (see Chapter 1, Figure 1.21). In most cases, however, the line positions will remain symmetrically disposed about the midpoint of the multiplet (Figures 1.21 and 1.22).
- By chance, two inequivalent neighbouring groups may have exactly the same coupling constant to the nucleus under observation. For example, a triplet will normally indicate a group of two equivalent spin ½ neighbours. However, two inequivalent spin ½ neighbours might show exactly the same coupling constant to the resonance we are looking at, giving the appearance of a triplet, when in fact, the resonance is a doublet of doublets.

Differentiation of chemical shifts and scalar couplings

The two most important effects in the NMR spectrum, from the point of view of structure determination, are the chemical shift and scalar (J) couplings. The chemical shift results in a separate multiplet for each chemically different group of NMR active nuclei in the molecule, i.e. each group of chemically different nuclei will have its own resonance position determined by the chemical nature of its surroundings. If we integrate each resonance in the spectrum, we can determine the relative, but not the absolute, number of nuclei responsible for each chemical shift. The structure within each multiplet results from scalar coupling and is determined by the number and arrangement of neighbouring groups of spins. Fortunately, it is a simple matter to distinguish chemical shifts from scalar couplings in most cases; the multiplet structure arising from scalar coupling in a first-order multiplet and in many second-order patterns is symmetrical about the midpoint of the multiplet, whilst the chemical shift places resonances according to the chemical nature of the groups giving rise to each resonance. Thus, there will be no pattern—*i.e. no symmetry*—in the distribution of the chemical shifts. Using the symmetry present in a scalar-coupled multiplet, we can then easily distinguish scalar couplings from chemical shifts in most cases and hence arrive at the number of chemically distinct groups of NMR active nuclei in the molecule. Occasionally an ambiguity might arise; for example four lines of equal intensity might be due to two doublets, a doublet of doublets, or coupling to a spin ½ nucleus. Fortunately, these situations can usually be quickly resolved by inspecting the other multiplets present and/or applying common sense—see 'Resolving ambiguities' later.

Figure 2.1 shows the ^{19}F NMR spectrum of a compound that contains fluorine, phosphorus, and alkyl groups, $P_xF_yR_z$. The spectrum contains ten lines; however, we notice that six of the lines give a symmetrical pattern, recognizable as a doublet of triplets, about a chemical shift of −88.6 ppm, $\delta(F^A)$. The remaining four lines also give a symmetrical pattern, a doublet of doublets, around 4.8 ppm, $\delta(F^B)$. Since we can identify two sets of symmetrical resonances, we know that the compound must contain two chemically different types of fluorine atoms, F^A and F^B. We can use the intensities of the multiplets to tell us the relative number of fluorines in each group. The total intensity of the four-line pattern is twice that of the six-line, therefore there must be twice as many fluorine atoms in the group B as in group A. However, although we know the ratio of F^A to F^B, we do not know the number of fluorines in each group; we must use the coupling pattern produced in the resonance of the neighbouring groups to work this out.

Determination of couplings and coupling patterns

The integration we did above told us that there were twice as many fluorines in group B as in group A but not how many fluorines are in each group. To find this out we use the coupling patterns. First, we need to identify which neighbouring group is responsible for each splitting pattern. To do this we use the fact that scalar coupling constants are reciprocal, that is the magnitude of the coupling between two nuclei A and B must be the same in the multiplet of A as it is in B. This allows us easily to determine which resonances come from mutually coupled groups of nuclei;

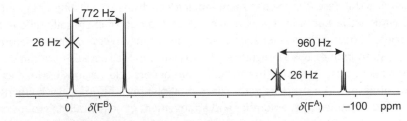

Figure 2.1 Coupling patterns are symmetrical about the midpoint and so are easily distinguished from chemical shifts. In the ^{19}F NMR spectrum above the doublet of doublets is thus easily distinguished from the doublet of triplets.

Source: From data in R. Schmutzler (1965), *Angew. Chem. Int. Ed.*, **4**, 496. Chemical shifts have been referenced to CCl$_3$F and the current standard of lower frequency/upfield shifts are negative.

we simply measure the coupling constants and look for two that are the same in different multiplets. We must, of course, remember that in the NMR spectra of inorganic compounds the coupling partner may be a nucleus of a different element—a heteronuclear coupling—and so its resonance appears in a different NMR spectrum.

Looking at the ^{19}F NMR spectrum of Figure 2.1 we notice that the separation of adjacent lines in each branch of the doublet of triplets is 26 Hz and is the same as the separation of the lines in each branch of the doublet of doublets. Immediately we can see that the 26 Hz coupling relates the two multiplets and so must be the coupling $J(F^AF^B)$. Because group F^A splits the resonance of its neighbour, group F^B, into a doublet, we know that group F^A must contain one spin ½; similarly, because group F^B splits group F^A into a triplet, group F^B must contain 2 spin ½ nuclei, i.e. there is one fluorine of type A and two of type B.

We also notice that the intensities of the first two lines of the resonance of F^A are in the ratio 1:2, confirming that the coupling is a triplet, i.e. due to a neighbouring group containing two equivalent spin ½. In fact, the separation of the first two lines of a first-order multiplet must correspond to a coupling constant and the relative intensity of the outside line to the next line must equal the number of coupling partners responsible for this coupling. Remember, however, that the separation of any other two lines in a multiplet will not necessarily be a coupling constant. These are useful rules when the coupling pattern is not immediately obvious; however, it does not hold in coupling in second-order systems (*vide* Chapter 1, section 1.10).

The large doublet coupling in the resonance centred at −88.6 ppm is 960 Hz, whilst in the resonance at 4.8 ppm it is 772 Hz. Clearly these couplings do not link the two multiplets in the ^{19}F NMR spectrum; these must be due to coupling to a heteronucleus. Since the coupling partner is clearly not in the ^{19}F NMR spectrum, we might guess that it is phosphorus, but must check the coupling constants in the ^{19}F multiplets against those in the ^{31}P NMR spectrum to be sure.

The ^{31}P{^{1}H} NMR spectrum (Figure 2.2), shows several lines around 8 ppm— remember that nuclei in braces, { }, have been decoupled, so couplings to protons are removed. The multiplet structure is less clear than that in the ^{19}F NMR spectrum (Figure 2.1); however, the whole pattern is symmetrical, indicating that it arises from scalar couplings on a single phosphorus chemical shift; i.e. there is only one type of phosphorus in the molecule. We know that the separation of the first two lines must equal a coupling constant, here 772 Hz, and that the ratio of the first two lines, 1:2, tells us the number of spins in the neighbouring group responsible for the coupling. We deduce that the 772 Hz coupling is most likely to a group of two identical spin ½ nuclei. Inspection of the ^{19}F NMR spectrum reveals that a 772 Hz coupling is also associated with the resonance of F^B so this group of fluorines must be the coupling partners.

By looking for this coupling recurring in the ^{31}P multiplet we can identify two 1:2:1 triplets: lines 1, 2, and 4 and 3, 5, and 6. Having identified the triplets, the multiplet structure is seen to be a doublet (960 Hz) of triplets (772 Hz), the 960 Hz coupling being due to the other fluorine, group F^A. Because the phosphorus group causes doublet splittings of the fluorine resonances, we know there is only one phosphorus present.

Practical tip: use a sharp pencil and tick marks on a piece of paper to compare coupling constants if a numerical print out is not available. Do not use a ruler; the latter is not sufficiently precise.

If comparing between spectra, make sure these are on the same Hz scale.

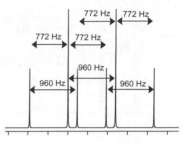

Figure 2.2 The ^{31}P{^{1}H} NMR spectrum above shows a doublet of triplets. Note that the spectrum is symmetrical about the midpoint, confirming that all lines belong to a single multiplet.

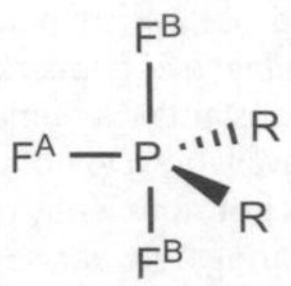

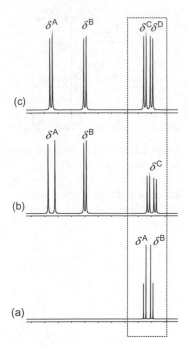

Figure 2.3 Analysis of the ^{19}F and ^{31}P NMR spectra helps us to deduce the structure of PF_3R_2. Other techniques such as mass spectroscopy and microanalysis must also be used to characterize the compound conclusively.

Combining all this information, we can deduce that the compound contains two types of fluorine, F^A and F^B in the ratio 1:2, that the fluorines are close enough to each other to be mutually coupled, and that both types of fluorine see coupling to another spin ½ nucleus, P. The splitting each group causes in its neighbours tells us that there is one of type F^A, two of type F^B, and one phosphorus. (It is also possible that the molecule might be a polymer of some sort containing isolated, but identical $\{P(F^A)(F^B)_2\}$ units.)

The common oxidation states of phosphorus are III and V; since we know the phosphorus has alkyl substituents as well as the three fluorines, we must have a P(V) compound and we might guess that there are two alkyl groups, i.e. the molecule is PF_3R_2 (Figure 2.3). The nature of the alkyl groups could be determined by a combination of ^{1}H and ^{13}C NMR; however, it is impractical to use couplings to ^{13}C in the ^{31}P or ^{19}F NMR spectra to determine the number of alkyl groups since ^{13}C is only 1% abundant—the multiplets, due to coupling to ^{13}C, will appear as satellites on the main peak and will contain only 1% of the intensity in the spectrum (see Chapter 1, section 1.9). The type and number of alkyl groups would usually be determined another way, for example by microanalysis, whilst mass spectroscopy could be used to determine the molecular mass (and hence whether the molecule is a monomer or polymer).

Resolving ambiguities

We noted earlier that ambiguities can arise; for example, four lines of equal intensity might be due to two doublets, a doublet of doublets, or coupling to a spin ³⁄₂ nucleus. In the real world, we will most likely know if a spin ³⁄₂ nucleus is present in the molecule. In the case we do not, we note that the coupling pattern arising from a spin ³⁄₂ neighbour is four equally spaced lines of equal intensity, a situation that is highly unlikely to arise from, say coupling to two inequivalent spin ½ nuclei (a doublet of doublets) or from two different nuclei each coupling to a spin ½ neighbour.

Looking at the multiplet structure, intensities, and number of the coupling partners in the spectrum will usually resolve ambiguities in spin ½ spin systems. For example, the upfield region of the spectra in Figure 2.4 each show a symmetrical four-line pattern (dotted box). An **AB** quartet (Figure 2.4a) arises from two mutually coupled spins, δ_A and δ_B, with similar chemical shifts and is easily identified by the roofing that occurs and the absence of any other coupling partners in the spectrum. The true doublet of doublets (Figure 2.4b) arising from δ_C coupling to two inequivalent spin ½ nuclei is readily identified both by the presence of two coupling partners elsewhere in the spectrum, δ_A and δ_B, having different J values $J(AC) \neq J(BC)$, and by its intensity relative to its coupling partners, while the rare case of two groups, δ_C and δ_D, having very similar chemical shifts *and* coincidentally having the same value coupling constant $J(AC) = J(BD)$ to one spin ½ neighbour each, δ_A and δ_B, *and* not coupling to each other or any additional spins (Figure 2.4c) can be readily distinguished from the doublet of doublets, since: (1) in this case both coupling constants in the neighbours have the same value, while the true doublet of doublets requires different J values for

Figure 2.4 Ambiguities can sometimes arise, for example (a) an AB quartet; (b) coupling to two inequivalent neighbours giving a doublet of doublets; and (c) two doublets arising from two spins with similar chemical shifts, each having one spin ½ neighbour and by chance having the same J value to their respective neighbours. Each gives a symmetrical four-line pattern (dotted box). Inspection of the rest of the spectrum will usually readily distinguish these possibilities.

the two neighbours; and (2) the relative intensities of the various resonances are different.

Experimentally these cases can be distinguished by recording a 2-D spectrum, or by a selective decoupling experiment, or by recording the spectrum at two different field strengths, since the coupling constants, J, do not vary with magnetic field, whereas the resonance frequency does.

Satellites

Satellites occur as a result of scalar coupling when the coupling partner has low natural abundance. Since these contain all the coupling information, satellites can also be used to obtain structural information from an NMR spectrum.

Example

Reaction of *cis*-[PtCl$_2$(PEt$_3$)$_2$] with SnCl$_2$ affords a new complex of empirical formula $C_{12}H_{30}Cl_4P_2PtSn$. By analysing the $^{31}P\{^1H\}$, ^{119}Sn, and ^{195}Pt NMR spectra shown in Figure 2.5 we can determine the structure of the new complex.

One symmetrical multiplet is seen in the NMR spectrum of each nuclide; there is one chemical shift in each spectrum, i.e. one type of phosphorus, one type of tin, and one type of platinum in the complex. Each spectrum shows a multiplet with satellites; we can tell by the intensity ratio of the outlying peaks to the central peaks that these must be satellites.

We know that platinum has one NMR active isotope, ^{195}Pt, and that this is 33.8% abundant. Tin has three spin ½ isotopes: ^{115}Sn (0.4%), ^{117}Sn (7.6%), and ^{119}Sn (8.4%) (Table 1.2). The satellites must be due to coupling to these nuclides.

Consider first the NMR spectrum of ^{119}Sn. The central resonances are a 1:2:1 triplet, so tin most likely has a neighbouring group containing two equivalent spin ½ nuclei. To find the coupling partner, we look for a coupling of the same magnitude, 237 Hz, in the multiplets in the other spectra. The coupling matches that labelled $J(^{31}PA)$ in the $^{31}P\{^1H\}$ NMR spectrum, so we assign this coupling as $J(^{119}Sn^{31}P)$. Since phosphorus splits tin into a 1:2:1 triplet, we know that, most likely, there are two equivalent phosphorus ligands, i.e. there are two PEt$_3$ ligands, and these are mutually *trans*, in contrast to the *cis* arrangement in the starting complex. The large coupling in the ^{119}Sn spectrum, 28954 Hz, matches the coupling labelled $J(^{195}PtB)$ so is assigned to the ^{195}Pt satellites, $J(^{119}Sn^{195}Pt)$ and is a doublet. That this is coupling to ^{195}Pt is confirmed by the intensity ratio, 1:4, which corresponds with the natural abundance of ^{195}Pt. Since platinum splits tin into a doublet, there must be one platinum in the complex; already we know the complex contains one platinum atom, and two phosphorus-containing ligands, as well as some tin.

Looking at the ^{195}Pt spectrum, we see that the central resonances are a 1:2:1 triplet, labelled $J(^{195}PtC)$ and the value of the coupling constant, 2042 Hz, equals the coupling to the major satellites in the $^{31}P\{^1H\}$ NMR spectrum, so we assign this coupling to $J(^{195}Pt\,^{31}P)$; phosphorus causes a triplet splitting on platinum confirming there are two, equivalent phosphine ligands in the complex. The

We can work out the proportion each isotopologue makes to the spectrum using simple probability—the probability of having a given isotope present of a given element is its percentage natural abundance. If we want a particular isotope of tin and a particular isotope of platinum, then we multiply the percentage natural abundances together.

For example, the probability that the platinum is ^{195}Pt is 33.8%, its percentage natural abundance. If we also want the tin to be ^{119}Sn, we multiply the ^{195}Pt percentage abundance by the percentage natural abundance of $^{119}Sn = 33.8 * 8.6\% = 2.9\%$. Remember this gives the *total* intensity of the $^{195}Pt^{119}Sn$; isotopologue satellite peaks and must be divided between the lines in the satellites.

The intensities of the ^{117}Sn and ^{115}Sn satellites can be calculated following the same procedure.

The chances that a given phosphorus is adjacent to two ^{119}Sn nuclei is $8.6\% * 8.6\% = 0.7\%$. This is so small it can be ignored. Thus, we would expect to see a doublet for the ^{119}Sn satellites on each phosphorus resonance. The intensity of each line in the doublet is 17.2% divided by $2 = 8.6\%$.

satellites look complex until we remember that tin has three spin ½ isotopes, so we are expecting three sets of satellites, each of which is a doublet of triplets (coupling to one of the tin isotopes–doublet, $J(^{195}Pt^{11X}Sn)$–and to the group of two phosphorus ligands–triplet, $J(^{195}Pt^{31}P)$). The very similar gyromagnetic ratios of the tin isotopes, and hence similarity of the $J(^{195}Pt^{X}Sn)$ coupling constants means the satellites overlap, but we can still deduce that one tin is present in the molecule.

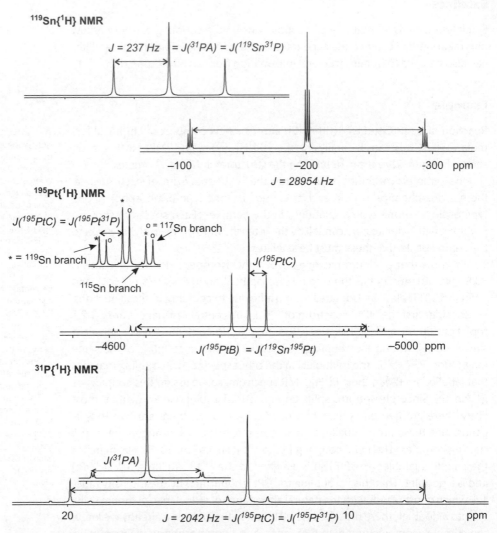

Figure 2.5 $^{31}P\{^{1}H\}$, $^{119}Sn\{^{1}H\}$ and $^{195}Pt\{^{1}H\}$ NMR spectra of the new complex formed on reaction of *cis*-$[PtCl_2(PEt_3)_2]$ with $SnCl_2$. Main ^{119}Sn and ^{195}Pt spectra are on the same scale. ^{119}Sn expansion is at ½ scale of the ^{31}P expansion; the ^{195}Pt expansion is $1/15$ scale of the ^{31}P expansion; the ^{31}P expansion is $4/3$ scale of the full ^{31}P spectrum on the Hz scale. NB assignment of the coupling partners should be made by comparing the J values. The central line of the ^{115}Sn branch is obscured by the more intense ^{117}Sn and ^{119}Sn lines.

Source: From data in O. Starzewski and P. S. Pregosin (1982), in *Catalytic Aspects of Metal Phosphine Complexes*, ACS, **196**, 23–41.

The central phosphorus resonance is a singlet with satellites due to coupling to tin, $J(^{31}PA) = J(^{119}SnX)$, and platinum, $J(^{31}PB) = J(^{195}PtQ)$. Both are doublets, confirming that there is one tin and one platinum in the complex.

This completes our analysis of the three NMR spectra. Now we have to assemble the groups into a complex. We have one tin centre, one platinum, and two equivalent phosphines. We can reasonably assume this must be a platinum complex, and if the oxidation state hasn't changed it will be Pt(II), so we expect a square planar geometry. One coordination site is occupied by Sn and the two phosphines must be *trans* to each other in order to be equivalent. Microanalysis told us there are four chlorines in the complex. One occupies the fourth coordination site on Pt and the other three are around the tin, **1**.

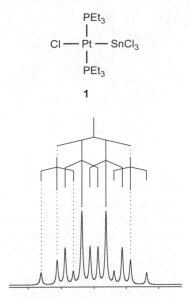

Analysis of overlapping first-order multiplets

Sometimes overlaps due to similar coupling constants can mean the multiplet structure is not obvious (Figure 2.6). We can use the fact that the separation of the first two lines in a first-order multiplet corresponds to a coupling constant and the intensity ratio of the two lines tells us the type of multiplet—we focus here on multiplets due to coupling to spin ½ nuclei. Although the multiplet structure in Figure 2.6 is not immediately obvious, the intensity ratio of the first two lines is 1:2, so this must be the start of a triplet and we can measure the coupling constant.

One way to search for coupling constants, if we have a printout of the spectrum, is to use tick marks on paper. We can then slide the paper across a multiplet to find all lines that 'belong together'; in this example, we are looking for the three lines of the triplet. Having found all three lines, we can draw a single line at the midpoint of the triplet. We now repeat the process from the other end and finally across the middle of the multiplet to find any other instances of our triplet, replacing each one with a single line at the midpoint. We can now see that we have four equally spaced lines in the ratio 1:3:3:1, i.e. the multiplet is a quartet of triplets. Our resonance has two neighbouring groups, one containing two equivalent spin ½ neighbours, the other three. Notice also that we have, in effect, constructed a stick diagram of the multiplet.

Example

In this example we see how to disentangle the structure of an overlapping multiplet.

Reaction of PF$_3$ with water affords **A**. The ^{1}H, ^{19}F and ^{31}P NMR spectra of **A** are shown in Figure 2.7. The ^{1}H NMR spectrum is one symmetrical multiplet, so there is one type of H present. N.B. We do not yet know how many protons of this type are present. The coupling pattern is a doublet of 1:2:1 triplets, so the proton has two neighbouring groups, one containing one spin ½ nucleus, the other containing two equivalent spin ½ nuclei. N.B. At this stage we do not know what elements these nuclei are.

Figure 2.6 Overlapping multiplets frequently occur; first-order multiplets can usually be disentangled following the procedure in the text.

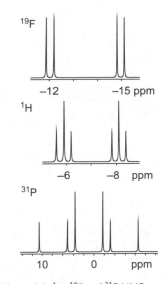

Figure 2.7 ^{1}H, ^{19}F and ^{31}P NMR spectra of **A**. ^{31}P spectrum is shown at half scale.

Source: From data in L. F. Centofani and R. W. Parry (1968), *Inorg. Chem.*, **7**, 1005.

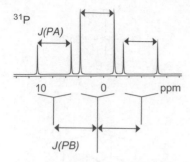

Figure 2.8 Once we have identified a repeating coupling in a multiplet, we can replace it by a single line at the centre of each branch. Here we replace the repeating doublet structure arising from coupling to nucleus A by a single line at the centre of each branch, allowing the triplet due to coupling to the two spin ½ s in group B to be clearly seen. In effect we generate the stick diagram for the spin system.

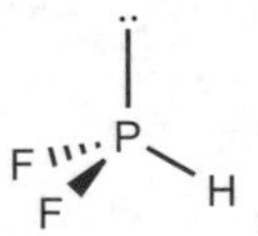

B

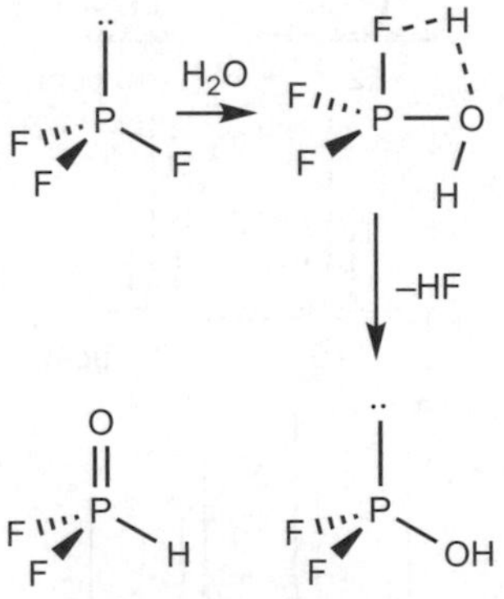

Scheme 1 The reaction of PF_3 with water.

The ^{19}F NMR spectrum also shows one symmetrical multiplet, so there is one fluorine chemical shift, i.e. one type of F is present. The coupling pattern is a doublet of doublets, so the fluorine has two neighbouring groups, each containing one spin ½ nucleus.

The ^{31}P NMR spectrum is also one symmetrical multiplet, so there is one phosphorus chemical shift, i.e. one type of P is present. The coupling pattern is not immediately obvious. However, we know that the separation of the first two lines is a coupling constant. And the intensity ratio of these two lines, 1:1, tells us we are looking at a doublet. Using tick marks on paper, we can find where this doublet repeats in the multiplet and replace each repeat with a single line at its centre (Figure 2.8). Immediately it is apparent that we have a doublet of 1:2:1 triplets; the phosphorus, therefore, has two neighbouring groups, one containing one spin ½ nucleus, the other containing two equivalent spin ½ nuclei. We also now know which lines in the spectrum to use to measure the coupling constants.

We know that if group **X** couples to group **Y**, $J(\mathbf{XY})$ must be the same in both multiplets. Again, we can use tick marks on paper to compare the coupling constants but must remember to correct for different scales; for example in Figure 2.7 the ^{31}P NMR spectrum is shown at half the scale of the others, so we must double any distance measured in the ^{31}P spectrum before comparing it with a distance in the ^{1}H or ^{19}F spectrum. Doing this, we find the triplet in the ^{31}P corresponds with the large doublet in the ^{19}F NMR spectrum; fluorine couples to phosphorus and splits the phosphorus resonance into a triplet, so the fluorine group must contain two fluorines. Similarly, we find that fluorine splits the proton into a 1:2:1 triplet, confirming there are two fluorines in the group. On the other hand, we find that the phosphorus and proton groups split their neighbours into doublets—there is only one phosphorus and one proton present. N.B. Each statement above contains different information, so is required in the logical sequence to assign the NMR spectra. With practice, you will perform many of these steps subconsciously, but at first it is important that you do each one *explicitly*.

We now apply our chemical knowledge to assemble the H, P, and two Fs into a molecule. The common oxidation states of phosphorus are P(III) and P(V). If we put the phosphorus in the centre of the molecule and form bonds to the H and 2Fs, we end up with structure **B**. Although this is a reasonable structure that fits the topology revealed by NMR data, the values of the coupling constants and chemical shifts are inconsistent with this structure and it is difficult to see how this could be formed by the reaction of PF_3 with water; **A** is actually a phosphine oxide, formed as shown in Scheme 1.

The method can be easily modified to deal with quadrupolar nuclei; the intensity ratio of the first two lines tells us the number of spins in the coupling group and the number of lines in the sub-spectrum (found using tick marks on paper) tells us the value of **I**.

2.3 Summary

- When analysing an NMR spectrum, consider the groups of spins present, not individual spins.

- Look for symmetrical patterns to identify the chemical shifts (the number of chemically different groups of spins of the nuclide/element observed) present in the molecule.

- Coupling is caused by neighbouring groups of spins. The patterns tell us about the membership of the neighbouring group, not about the membership of the group whose resonance we are looking at.

- Use the coupling pattern/multiplicity and coupling constants to identify the neighbouring groups and number of spins in each neighbouring group responsible for each coupling pattern.

- Coupling may be homonuclear—the coupling partner is of the same nuclide, so its resonance appears in the same spectrum—or heteronuclear—the coupling partner is of a different nuclide, so its resonance appears in 'the other' NMR spectrum. The rules for interpreting homo- and heteronuclear couplings are the same.

2.4 Exercises

General comment

The aim of these exercises is to give practice in the analysis and interpretation of NMR spectra, not to determine the structure of the molecule. The reader is encouraged to analyse each spectrum as fully as possible.

1. Figure 2.9 shows the $^{31}P\{^1H\}$ NMR spectra of the dimeric phosphinidene $\{M(\mu\text{-}PAr)\}_2$ **1**, **2**, **3** and the $^{207}Pb\{^1H\}$ spectrum of **2**. Account for the all couplings observed. Suggest a reason for the inability to observe the ^{73}Ge NMR spectrum.

2. Reaction of $[PtH(dtbpx)(MeOH)][OTf]_2$ **A** with ^{13}CO affords **B**. The NMR data for **A** and **B** are given in Table 2.1. The ^1H and $^{31}P\{^1H\}$ NMR spectra are shown in Figure 2.10.

   a. Use the fact that first-order coupled multiplets are symmetrical to assign the ^{195}Pt satellites to their corresponding central peaks in the NMR spectra in Figure 2.10.

   b. Use the magnitudes of the various coupling constants to determine the stereochemistry around the Pt centre.

   c. Suggest a reason for the roofing observed in the $^{31}P\{^1H\}$ NMR spectrum of **B**.

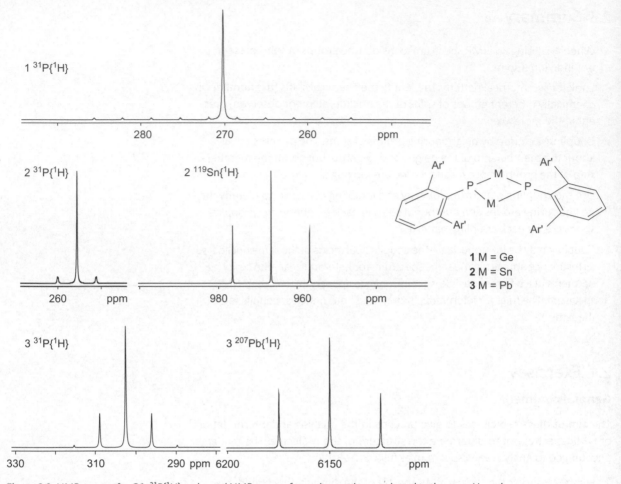

Figure 2.9 NMR spectra for Q1; $^{31}P\{^{1}H\}$ and metal NMR spectra for each complex are plotted to the same Hz scale.

3. Advanced question. Interpret fully the spectra in Figure 2.10 and Table 2.1 and hence confirm the identities of **A** and **B**.

4. Advanced question. The three spectra shown in Figure 2.11 each show a symmetrical six-line pattern in the upfield region, yet each arises from a different spin system.

The first spin system comprises: group **A** containing two equivalent spins and two groups, **B** and **C**, each containing one spin ½ nucleus. **B** and **C** have similar chemical shifts and do not couple to each other but do couple to group **A** with, accidentally, the same coupling constant $J(\mathbf{AB}) = J(\mathbf{AC})$. **B** and **C** are of the same element, but **A** is a heteronucleus.

The second spin system comprises: group **A** containing one spin ½ nucleus and two groups, **B** and **C**, also each containing one spin ½ nucleus. **B** and **C** have similar chemical shifts and couple to each other and to group **A**; all the coupling constants are, accidentally, the same. **A** is a heteronucleus.

Finally, the third spin system comprises: groups **A** and **B** each containing one spin ½ nucleus and group **C** containing two spin ½ nuclei. **B** and **C** have dissimilar

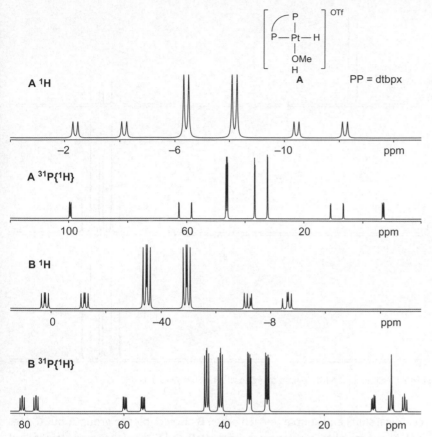

Figure 2.10 NMR spectra for Q2 and 3 recorded at $^1H = 100$ MHz; selective proton decoupling has been used to remove couplings to phenyl protons. The aromatic region of the ^1H NMR spectra are not shown. A slight second-order distortion can be seen in the upfield satellites of the ^1H and $^{31}P\{^1H\}$ NMR spectra of A. The latter also contains accidental overlapping of lines the upfield satellites of the two different phosphoruses in B.

Table 2.1 Selected NMR data for the complexes

	δ/ppm[a]	J(P^1–	J(P^2–	J(Pt–[b]
Compound A				
P^1	46.2 d	–	6	4312
P^2	34.6 d	6	–	2095
H	−7.3 dd	18	176	805
Compound B				
P^1	42.2 dd	–	19	2958
P^2	33.2 dd	19	–	2011
H[c]	−4.2 ddd	15	145	740
CO	179.3 dd	113	8	1299

[a] All spectra recorded with ^1H decoupling, where given couplings to ^1H were determined from the ^1H NMR spectrum.

[b] ^{195}Pt satellites are seen on all resonances.

[c] J(HC) not determined. d = doublet.

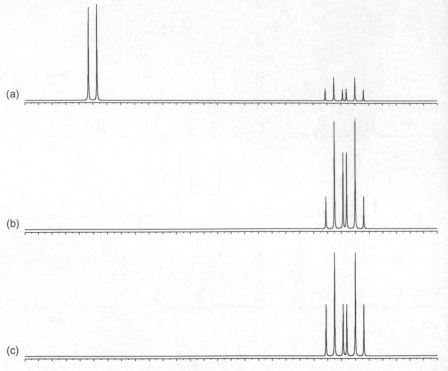

Figure 2.11 Simulated NMR spectra of groups B and C; for Q4 see text.

chemical shifts and couple to each other. **B** also couples to group **A** but **C** does not. All the coupling constants are different. **B** and **C** are of the same element but **A** is a heteronucleus.

Using the workflow in section 2.2 ('Interpretation of an NMR spectrum') and paying particular attention to the caveats and information in the side bar box, explain which spectrum corresponds to which spin system and how these can be differentiated.

Hint: Start by constructing stick diagrams for each resonance.

3 Factors influencing the chemical shift and coupling constants

3.1 Introduction

In Chapters 1 and 2 we looked at some essential theory of NMR, how we can predict the NMR spectrum and how to analyse and use chemical shift and coupling information to deduce the structure of the compound. In the interpretation of an NMR spectrum we are, of course, not interested solely in the structure of the molecule, but also in obtaining as much information as possible, for example about stereochemistry, oxidation state, co-ordination number, bond angles, etc. This chapter surveys the chemical factors that influence the chemical shift and the magnitude of coupling constants.

3.2 General trends in the chemical shift

Several factors, for example co-ordination number, oxidation state, and geometry, influence the shielding of the nucleus and hence the chemical shift in inorganic complexes. Although it is possible, by carefully chosen examples where a single factor dominates the shielding, to establish trends in the chemical shift, it is difficult to establish general rules for the chemical shifts of nuclei in inorganic compounds, since many interdependent, often competing, factors are involved. Shifts can be very large—up to 10,000 (or more) ppm in metal complexes, which is perhaps unsurprising given the often strong interactions of the metal electrons with the ligands.

Traditionally, chemical shifts to higher ppm are described as 'downfield' or 'deshielded' and shifts to lower and negative ppm as 'upfield' or 'shielded'. More recent texts, however, describe shifts as to 'higher' or 'lower' frequency, respectively. This reflects the move from continuous wave NMR, in which the spectrum is recorded by changing the magnet field strength, to Fourier transform NMR, in which spectra are recorded in the time domain—see Chapter 4—and relates the terminology of the ppm scale directly to the separation of the energy levels; a larger separation increases the resonance frequency and gives an increase in chemical shift. It is worth reiterating at this point that an increase in shielding results in a decrease in the chemical shift, δ, i.e. in a shift upfield, or to *less positive/more negative δ values*.

Factors influencing the chemical shift—the diamagnetic and paramagnetic terms

Chapter 1, section 1.6 introduced the concept of diamagnetic and paramagnetic contributions to shielding. The diamagnetic term is generated by ground state electrons, is dominant for light atoms, and is generally insensitive to changes in the chemical environment. As a result, the chemical shift ranges of light atoms such as protons and lithium are small. The paramagnetic term arises from electrons mixed in from excited electronic states and so provides important chemical information about the electronic structure and bonding in molecules, for example the energy difference between the frontier orbitals (HOMO and LUMO). The paramagnetic term is often dominant for nuclei heavier than Li and gives much larger shifts, which can be to higher or lower frequency. We must be careful not to confuse the paramagnetic contribution to the chemical shift with the effects of paramagnetism, for example from paramagnetic impurities.

Factors influencing the chemical shift—geometry

We have already seen that geometric factors can make otherwise equivalent nuclei distinct, for example the axial and equatorial fluorines in the AsF_3R_2 example in Chapter 1, section 1.6. Other such geometric factors might be *cis* versus *trans* or *mer* versus *fac* isomerism in a complex or the overall symmetry present or absent in the compound (Figures 3.1 and 3.2). It is therefore important that we can recognize when geometry makes, say, the ligands in a complex different.

A useful starting point is to look at the group *trans* to the ligand of interest; thus in *fac*-Mo(CO)$_3$(PPh$_3$)$_3$ each triphenylphosphine ligand is *trans* to a CO, whereas in the *mer* isomer two PPh$_3$s are mutually *trans* whilst the third is *trans* to a CO ligand; there will be one phosphorus chemical shift in the ^{31}P{^{1}H} NMR spectrum of *fac*-Mo(CO)$_3$(PPh$_3$)$_3$ but two in the spectrum of the *mer* isomer in the ratio 2:1.

Similarly, one triphenylphosphine ligand in Wilkinson's catalyst (Figure 3.2), is *trans* to chloride and is clearly different to the two mutually *trans* triphenylphosphines. Thus, two phosphorus resonances are seen and coupling to rhodium and between the dissimilar phosphines is seen.

In the examples above, it was easy to recognize the spatial arrangement that makes otherwise identical ligands different; in other structures more careful consideration is required. For example, in Fe$_3$(CO)$_{12}$ (Figure 3.3), we can easily see that the two bridging carbonyls, CO^B, will be distinct from the ten terminal CO ligands. Deciding which terminal COs are the same and which are inequivalent, however, requires thought. The terminal carbonyls on Fe^A must be different from those on Fe^B, since the iron atoms are clearly different. However, two of the

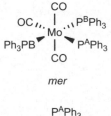

Figure 3.1 Geometric isomerism can make otherwise identical species different. Here the triphenylphosphine ligands in the fac complex are equivalent but in the mer isomer there are two types of PPh$_3$ ligand differentiated by their geometrical position.

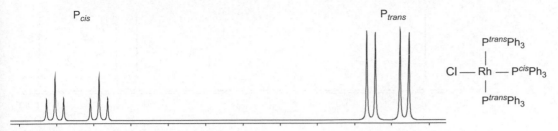

Figure 3.2 The differing ligands in the *trans* positions divide the sites occupied by the triphenylphosphine ligands in Wilkinson's catalyst into two distinct groups. In the ^{31}P{^{1}H} NMR spectrum the inequivalent groups of phosphorus nuclei couple to each other as well as to the rhodium.

carbonyls on Fe^A, (C^2O), lie in the Fe_3 plane, while two (C^1O) lie in symmetrically equivalent positions above and below the Fe_3 plane; thus C^1O and C^2O are distinct groups. In a similar way, the terminal CO ligands on Fe^B divide into two groups: C^3O lie *trans* to the Fe^A–Fe^B bond, whilst C^4O lie above and below the Fe_3 plane. Thus, there are four distinct terminal CO environments and one type of bridging CO in $Fe_3(CO)_{12}$.

Finally, some compounds are 'flexible', for example 5-coordinate complexes of transition metals and other molecules having trigonal bipyramidal geometry in which Berry pseudo-rotation which moves the ligands/substituents between the axial and equatorial sites, may occur. If this interchange is fast then a single resonance, at the average chemical shift, will be observed rather than separate resonances for the different sites. The NMR of such dynamic systems is discussed in Chapter 6.

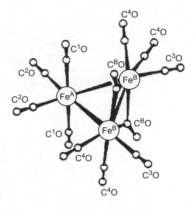

Figure 3.3 The spatial arrangement of the terminal carbonyl ligands in $Fe_3(CO)_{12}$ divides these ligands into four separate groups

Source: Adapted with permission from C. J. Wei and L. Dahl (1969), *J. Amer. Chem. Soc.*, **91**, 1531. Copyright © 1969 American Chemical Society.

Factors influencing the chemical shift—electronegativity, charge, and oxidation state

All of these factors affect the electron density around the nucleus and so affect the chemical shift. Shielding generally becomes smaller, i.e. the chemical shift increases, as the electron density at the nucleus decreases. Thus, electronegative substituents, positive charge, or an increase in oxidation state all (usually) result in downfield shifts.

The effect of substituent electronegativity can be seen in the shifts caused by halogen ligands in main group halide complexes—the shielding increases as the electronegativity of the halide decreases, thus shifts follow the electronegativity Cl > Br > I (Table 3.1). The opposite trend, however, may be observed for transition metals due to ligand field effects (Table 3.4).

The effects of electronegativity, charge, and oxidation state can also be seen, for example, in the ^{31}P chemical shift of some typical phosphorus compounds. In PPh_3 (P^{III}) the shift is about –6 ppm, whilst in $P(OPh)_3$ (more electronegative substituents) it is 128 ppm. In the oxide, $OPPh_3$ (P^V), it is around 28 ppm (increase in oxidation state), and in PPh_4^+ (P^V) the shift is 20.8 ppm (positive charge) (Figure 3.4).

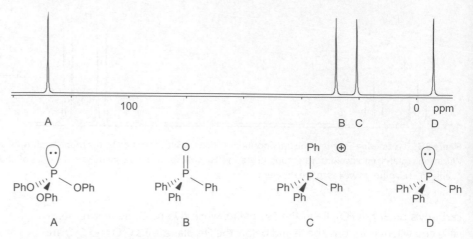

Figure 3.4 The chemical shift varies with substituents, oxidation state, and charge as seen here for some phosphorus compounds

Table 3.1 The shifts caused by halogen ligands usually follow the electronegativity $Cl^- > Br^- > I^-$

Compound	Chloride	Bromide	Iodide
AlX_4^-	103	80	−26.7
GaX_4^-	252	68	−502
InX_4^-	442	177	−569
SiX_4	−20	−94	−346
GeX_4	31	−311	−1081
SnX_4	−50	–	−1705
CdX_4^{2-}	417	333	261
$MeHgX$	−814	−915	−1097

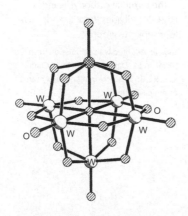

Figure 3.5 The molecular structure of $W_6O_{19}^{2-}$ contains three inequivalent sets of oxygen atoms. Unhatched circles are W. The apical tungsten (cross-hatched) can be substituted by a variety of other groups.

Source: Adapted from J. Errington et al. (1996), *J. Chem. Soc., Dalton Trans.*, 681. Reproduced by permission of The Royal Society of Chemistry. Copyright © 1996.

Factors influencing the chemical shift—co-ordination number

The effect of co-ordination number on the chemical shift is nicely illustrated in the ^{17}O NMR spectra of polyoxometallate anions (polyoxometallates are of interest as superacid catalysts, as precursors to metal oxides, and for the variety of metals and non-metals that can be incorporated into the metal core). A wide variety of heteroatoms can be included in the hexanuclear $[W_6O_{19}]^{2-}$ core (Figure 3.5 and Table 3.2).

Assignment of the ^{17}O NMR spectra of these compounds is straightforward, since the ^{17}O chemical shift is diagnostic of the co-ordination number, the resonance shifting upfield as oxygen co-ordination number increases, while the intensity of the ^{17}O resonance allows the number of each type of oxygen to be determined (Table 3.2).

Table 3.2 ^{17}O chemical shifts for some heteropolyanions based on $[W_6O_{19}]^{2-}$

Anion	M = O	W = O	W(μ₂–O)M	W(μ₂–O)W	μ₆–O
$[W_6O_{19}]^{2-}$		774	525	414	−80
$[(MeO)TiW_5O_{18}]^{3-}$		721,713	499	390,380	−58
$[(MeO)NbW_5O_{18}]^{2-}$		758,748	472	405,402	−71
$[(MeO)TaW_5O_{18}]^{2-}$		756,744	425	404,403	−68
$[VW_5O_{19}]^{3-}$	1224	737,734	565	399,392	−72
$[NbW_5O_{19}]^{3-}$	800	732,730	456	394,392	−67
$[TaW_5O_{19}]^{3-}$	666	739,736	426	397	−71

Source: From data in J. Errington et al. (1996), *J. Chem. Soc., Dalton Trans,* 681.

Factors influencing the chemical shift–co-ordination: effect on the ligands

Co-ordination shifts for a ligand are usually small—a few tens of ppm—and can be of either sign. For example, hydride ligands and the ligating carbon of alkyl ligands are usually shielded on co-ordination to a transition metal, whereas for strong π-acceptor ligands such as carbonyls and phosphines the ^{13}C and ^{17}O, and ^{31}P are usually deshielded on co-ordination, due to a paramagnetic contribution from the π-back bonding. For instance, Figure 3.6 plots the ^{13}CO chemical shift

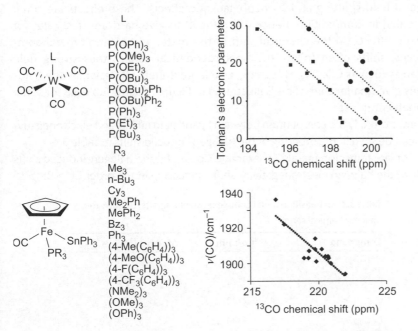

Figure 3.6 The chemical shift of the carbonyl carbon increases as the π-back donating power of the metal increases: ● CO_{trans} ■ CO_{cis}

Sources: Constructed from data in O. A. Gansow et al. (1971), *J. Amer. Chem. Soc.,* **93**, 5922; R. J. Randall et al. (1973), *J. Chem. Soc., Dalton Trans.,* 1027; L. Bodner et al. (1980), *Inorg. Chem.,* **19**, 1951; and R. M. Mampa et al. (2014), *Organometallics,* **33**, 3283.

against the Tolman electronic parameter of the phosphorus ligand or CO stretching frequency for two series of complexes, $W(CO)_5L$ and $CpFe(CO)(PR_3)(SnPh_3)$. The donor/acceptor ratio for phosphorus ligands increases in the order: phosphites < tri-arylphosphines < $P(NR_2)_3$ < tri-alkylphosphines; as the phosphine becomes more basic/electron-donating, the ^{13}CO chemical shift increases, illustrating this trend. We note, however, that separating steric effects from electronic effects and from differences in the σ-donation and π-acceptor properties of ligands is not always straightforward.

Other factors may also be important in the shielding, thus, in the parent carbonyl, $M(CO)_6$, the shift of the carbonyl carbon decreases from $\delta_C = 212.3$ to 204.1 to 191.9 ppm from M = Cr to Mo to W, although the IR stretching frequency is virtually unchanged, indicating little change in the π-back bonding. The variation in δ_C in this series of compounds has been attributed to an increasing diamagnetic contribution to the shielding and/or a decreasing paramagnetic contribution to the shielding from low lying, excited states as the triad is descended (3d to 4d to 5d metal).

Factors influencing the chemical shift–co-ordination: effect on transition metal

In addition to differences in steric effects and in the σ-donor and π-acceptor contributions of the ligands which can be challenging to disentangle, transition-metal chemical shifts are also influenced by both the magnitude of the ligand field splitting and by nephelauxetic effects. These effects are often (usually) in competition, hence it is difficult to give hard-and-fast rules for ligand effects on the chemical shift of the metal. In transition metal complexes, shifts often follow the 'spectrochemical series'; thus strong field ligands such as CO, PR_3, H^- usually give large upfield shifts, whilst the halogens give smaller shifts. This is illustrated in Figure 3.7 and Table 3.3 for some cobalt complexes.

Interestingly the competition between ligand field effects and electronegativity can be seen in the ordering of the halogen-induced shifts in Table 3.4.

For late transition metals, for example Co and Pt, the halogen-induced shifts follow the halogen electronegativity, shifts increasing in the order $Cl^- > Br^- > I^-$,

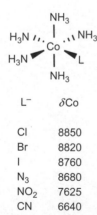

L^-	δCo
Cl	8850
Br	8820
I	8760
N_3	8680
NO_2	7625
CN	6640

Figure 3.7 The effect of the ligand on the chemical shift of the metal in transition metal complexes often follows the spectrochemical series

Source: From data in N. A. Matwiyoff et al. (1970), *Inorg. Chim. Acta*, **4**, 460.

Table 3.3 In transition metal complexes shielding often follows the 'spectrochemical series'

Compound	$\delta^{59}Co$/ppm	ΔE/cm^{-1} ($^1A_{1g} \rightarrow {}^1T_{1g}$)
$Co(H_2O)_6^{3+}$	15,100	16,500
$Co(NH_3)_6^{3+}$	8160	21,050
$Co(en)_3^{3+}$	7110	21,400
$Co(bipy)_3^{3+}$	6670	22,230
$Co(diars)_3^{3+}$	−100	23,200

Source: R. Goodfellow (1987), 'Group VIII Transition Metals', in *Multinuclear NMR*, ed. J. Mason, Plenum Press, New York.

Table 3.4 Ligand electronegativity and crystal field effects exert opposing influences on the metal chemical shift in halo complexes of transition-metals

Metal	Fluoride	Chloride	Bromide	Iodide
TiX_4	–	1162	1645	2440
NbX_6^-	−1490	0	731	–
Cp_2ZrX_2	–	−122	0	+126
WX_6	−1120	2050	–	–
$CpW(CO)_3X$	–	−1296	−1474	−1886
$Mn(CO)_5X$	–	−1005	−1160	−1485
$Co(NH_3)_5X$	–	8850	8820	8760
mer-$RhX_3(PMe_3)_3$	–	2202	1746	809

as we saw for main group halides, whilst for early transition metals, for example Sc, Ti, and Nb, shifts increase as the halide ligand crystal field strength $I^- < Br^- < Cl^-$ decreases due to the increasing importance of the paramagnetic term as the HOMO–LUMO gap decreases, a phenomenon termed 'inverse halogen dependence'. Clearly there is a balance between the paramagnetic contribution to the shift (downfield) from mixing in of low-lying, excited states and the diamagnetic effect (upfield shift) of the halogen electronegativity. Oxidation state can also affect this balance; compare the trends in shifts in the tungsten compounds WX_6 and $CpW(CO)_3X$, the high oxidation state compound showing 'inverse' whilst the low oxidation state complexes show 'normal' halogen dependence.

Factors influencing the chemical shift–paramagnetic centres and impurities

The magnetic moment of the electron is around one thousand times greater than a nuclear magnetic moment. The presence of unpaired electrons either in the molecule under study or as an impurity will have an impact on the NMR spectrum, since this introduces additional magnetic fields that lead to enhanced relaxation and to paramagnetic shifts of the resonance. The most widely used paramagnetic agents in NMR are lanthanide complexes. Lanthanide-induced shifts (LIS) and enhancement of spin–lattice relaxation have therefore received much attention.

Enhanced relaxation results in broadening of the NMR lines and follows from the Uncertainty Principle (equation 1.5), and, if extreme, can result in no observable NMR signal. Lanthanide-induced enhancement of spin–lattice relaxation rates may also 'wash out' the fine structure of multiplets due to J coupling in spectra when $1/T_1 > J$. This 'simplification' of NMR spectra is characteristic of paramagnetic complexes of both lanthanide and 3d-elements. Paramagnetic complexes such as β-diketonates of Cr^{2+}, Mn^{2+}, Fe^{2+}, Co^{2+} Ni^{2+}, and Cu^{2+}, or paramagnetic Gd(III) complexes may be deliberately added to a sample to speed up or level relaxation rates, for example where quantitative ^{31}P or ^{13}C spectra are required, and are used as contrast agents in magnetic resonance imaging (MRI)—see Chapter 4, section 4.5.

Lanthanide-induced shifts

The observed chemical shifts of the ligand in paramagnetic lanthanide complexes is the sum of three terms, the diamagnetic shift (δ_D, the shift in the absence of unpaired electron spin), the Fermi contact shift (δ_{FC}), and pseudo-contact shift (δ_{PC}) (equation 3.1).

Eq. 3.1 The chemical shifts of the ligands in lanthanide complexes depend on the diamagnetic shift δ_D, the Fermi contact shift, δ_{FC}, and pseudo-contact shift, δ_{PC}.

$$\delta_{LIS} = \delta_D + \delta_{FC} + \delta_{PC}$$

The Fermi contact shift arises from the interaction of the occupied electron shells of the ligand with the partially filled electron shell of the paramagnetic cation—this is sometimes expressed as delocalization of the unpaired electron(s) of the lanthanide cation onto the ligand—and depends both on the identity of the paramagnetic cation and on the nuclide observed in the ligand. δ_{PC} is a dipolar contribution from the through-space interaction of the nuclear and electron spins, and diminishes with distance as $1/r^3$. It should not be confused with δ_D, the shift in the absence of the paramagnetism. δ_{PC} is normally the dominant interaction, except for the nuclei of ligands directly bonded to the paramagnetic centre.

Paramagnetic shifts can greatly increase the chemical shift range, in protons by as much as 100 ppm or more. This can be useful in 'simplifying' complex spectra by 'stretching out' the chemical shift axis, thus separating overlapping resonances, and in separating the resonances of individual components in complex mixtures, for example in determining enantiomeric excess. Complexes used for this last purpose are known as chiral shift reagents.

Lanthanide shift reagents

In solution, lanthanide ions can form donor–acceptor encounter complexes with other compounds present in solution. If the lanthanide complex is chiral, diastereomeric encounter complexes, giving separate NMR resonances, form. The paramagnetic shift will increase the separation of these resonances, allowing enantiomeric excess of the donor to be determined. The most widely used paramagnetic shift agents are complexes of β-diketonates such as the camphor-derived complex of Figure 3.8. The presence of electron-withdrawing fluorinated groups in the β-diketonate ligand increases the affinity of the metal ion for hard Lewis bases such as nitrogen- and oxygen-containing compounds, allowing virtually any nitrogen- or oxygen-containing compound to be analysed. Lanthanide–silver reagents have been developed for soft Lewis bases such as alkenes, aromatics, phosphines, and halogenated compounds.

The increased difference between the chemical shifts of, for example, non-equivalent CH_2 protons due to LIS may also be used to remove second-order effects that arise when coupled spins have similar chemical shifts (cf.

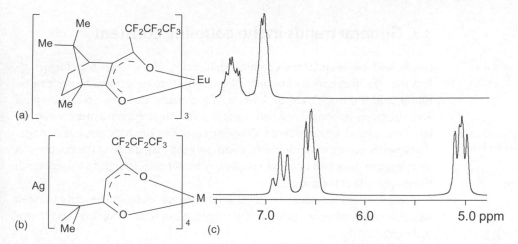

(a)

(b)

(c)

Figure 3.8 (a) Lanthanide complexes of camphor-based ligands can be used to separate the resonances of enantiomers in chiral mixtures containing hard Lewis acid donors; (b) Mixed Ln/Ag-fod complexes can be used with soft donors; (c) ^1H NMR spectra of triphenylphosphine in the absence (top) and presence of Ln(fod)$_3$/Ag(fod)$_2$.

Sources: From data in S. Sorenson and H. J. Jakobsen (1974), *Acta Chem. Scand.*, **A28**, 249; and T. J. Wenzel and R. E. Sievers (1982), *Anal. Chem.*, **54**, 1602.

Chapter 1, section 1.10), since the chemical shift difference (in Hz) will now considerably exceed the spin–spin coupling constants. Complexes of europium(III) are often chosen for this purpose, since these produce downfield shifts, in effect adding to the chemical shift. Paramagnetic broadening can become problematic on modern, high-field spectrometers; this can favour the use of samarium(III) or cerium(III) complexes at high field, even though these ions cause the smallest paramagnetic shifts of all the lanthanides, or the use of $^{13}C\{^1H\}$ rather than ^1H observation when recording NMR spectra at high magnetic field.

Temperature dependence of the paramagnetic shift: an NMR thermometer

The paramagnetic shift is temperature dependent, the Fermi contact term varying as $1/T$ and the pseudo-contact terms having a $1/T$ and $1/T^2$ dependence. This can be useful in, for example, in vivo temperature measurements since, over a biologically relevant temperature range, this reduces to a simple inverse relationship allowing paramagnetic complexes to be used as non-invasive temperature probes. Both lanthanide and transition metal complexes of macrocyclic ligands may be used for this purpose, with the choice of metal ion and ligand being dependent on the balance between the line broadening caused by the paramagnetic centre and the temperature sensitivity of the paramagnetic shift. Of course, binding of the metal ion by the ligand must be strong to prevent release of potentially toxic metal ions from the complex. Figure 3.9 illustrates the temperature dependence of a Tm(III) complex.

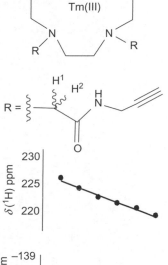

Figure 3.9 The temperature dependence of the paramagnetic shifts of transition metal and lanthanide complexes can be used to measure temperature: top, ring protons, bottom, H1 and H2. (1 and 2 superscripted)

Source: From data in M. Milne and R. H. E. Hudson (2011), *Chem. Commun*, **47**, 9194.

3.3 General trends in the coupling constant

Just as we have seen for the chemical shift, many, often competing, factors affect the size of coupling constants. Nevertheless, some general trends can be found, and the magnitude of the coupling constant can give useful chemical and structural information when variations in coupling constant between related complexes are considered. Coupling constants can have positive or negative signs; however, most chemists report only the magnitude of the coupling. A less negative value of a negative coupling constant corresponds to a decrease in the magnitude of the coupling.

Typical values for scalar coupling constants are given in Brevard's book—a valuable ready-reference guide to the magnitude of scalar couplings and chemical shifts.

Eq. 3.2 A reduced coupling constant, K, can be defined that removes the influence of the gyromagnetic ratios allowing comparison of coupling constants between different pairs of nuclei. K has units $NA^{-2}m^{-3}$.

$$K_{XY} = 4\pi^2 J_{XY} / h\gamma_X\gamma_Y$$

Factors influencing the coupling constant–gyromagnetic ratio

For a given pair of elements, the scalar coupling is proportional to the product of the gyromagnetic ratios of the nuclei provided all other factors remain the same. Thus, where two NMR active isotopes of a given element exist, for example ^{10}B and ^{11}B, the ratio of the coupling constants to a third nucleus will be as the ratio of the gyromagnetic ratios (Table 1.2). For example, in BH_4^-, $^1J(^{10}BH)$ is 27.0 Hz, whilst $^1J(^{11}BH)$ is 80.6 Hz (Figure 1.19). (The gyromagnetic ratios of ^{10}B and ^{11}B are 2.87 and 8.58 $\times10^7$ rad T^{-1} s^{-1}, respectively.) However, when we move from one element to another, other factors also change, for example the nature of the bonding; thus $^1J(PH)$ is *ca.* 200–500 Hz, whereas $^1J(PPt)$ is several thousand hertz despite the gyromagnetic ratio of the proton being roughly five times that of ^{195}Pt.

Factors influencing the coupling constant–periodicity

The magnitude of the reduced coupling constant (equation 3.2) increases down a group. For example, K_{EC} increases from 8.3 to 18.0 to 30.0 to 39.3 $\times10^{20}$ $NA^{-2}m^{-3}$ in the tetramethyl complexes of Group 14 metals $SiMe_4$, $GeMe_4$, $SnMe_4$, and $PbMe_4$. Note that the more familiar J varies (apparently randomly) from −50 to −19 to −340 to 250 Hz, illustrating the benefit of using the reduced coupling constant when comparing couplings and homologous series.

Two bond coupling constants generally increase as the intervening atom moves down a group. For example, $^2J(SnESn)$ in cyclic heterotristannanes (Figure 3.10), increases from 215 to 238 to 252 Hz as E descends the group from S to Se to Te. Note also the increasing shielding of the tin nuclei as the electronegativity of the substituents decreases.

As always, some care is needed if only the magnitude (but not the sign) of the coupling constant is reported, for example although $^2J(PH)$ in the series of olefin hydroformylation catalysts $HM(CO)(PPh_3)_3$, M = Co, Rh, Ir are reported to be in the order Co > Rh < Ir (Figure 3.11), it is probable that the coupling constant in the cobalt complex is negative.

E	δ(Sn) /ppm	^2J(SnSn) /Hz	δ(E) /ppm	^1K(SnE) (*10^{20}) /$NA^{-2}m^{-3}$
S	18	215	133	
Se	−43	238	−436	154
Te	−201	252	−991	236

Figure 3.10 $^2J(^{119}SnE^{117}Sn)$ in cyclic hetero-tristannanes increases as E descends the group from S to Se to Te

Source: From data in H. Lange et al. (2002), *J. Organomet. Chem.*, **660**, 43.

$$RCH_2 = CH_2 + CO + H_2 \xrightarrow{HM(CO)(PPh_3)_3} RCH_2CH_2CHO$$

	M	J(PH)/Hz
	Co	(−)50
	Rh	14
	Ir	42

Ph₃P — M with H (up), PPh₃, PPh₃, CO

Figure 3.11 The coupling constant through an intervening atom increases as the group is descended

Factors influencing the coupling constant–s-character in the bond

J coupling between nuclei is transmitted *via* the bonding electrons rather than directly through space. Clearly for coupling to occur the bonding electrons must be able to sense, or *contact*, the nucleus. Only s-electrons have a finite density at the nucleus, so it should come as no surprise that the amount of s-character in the valence hybrids used to form the chemical bond plays an important role in determining the strength of the coupling. Percentage s-character will be influenced by factors such as hybridization, co-ordination number, the electronegativity of substituents, interbond angles, the nature of the *trans*-ligand, oxidation state, etc.

Factors influencing the coupling constant–hybridization

The effect of changes in the formal hybridization of the nuclei on the coupling constant can be seen, for example, in PtC coupling constants which increase from 450–700 Hz in platinum alkyl complexes (sp^3-hybridized carbon) to 750–850 Hz in carbene complexes (sp^2 carbon) to 990–2250 Hz in terminal platinum carbonyls (sp carbon). This effect is also illustrated in the unusual complex shown in Figure 3.12, $(PMe_3)_2W(\equiv CCMe_3)(=CHCMe_3)(-CH_2Me_3)$, which contains an alkyl (sp^3 carbon), alkylidene (sp^2 carbon), and alkylidyne (sp carbon) ligand. The tungsten–carbon coupling constants are 80, 120, and 210 Hz, respectively.

Similarly, in rhodium carbonyl cluster compounds J(RhC) falls from *ca.* 70 Hz for terminal carbonyl ligands (sp) to 25 Hz for face-bridging ligands (sp^3) (Figure 3.13).

Factors influencing the coupling constant–co-ordination number

Increasing co-ordination number leads to a decrease in coupling constant; for example J(SiF) falls in the series SiF_4 (170 Hz) > $[SiF_5]^-$ (140 Hz) > $[SiF_6]^{2-}$ (108 Hz) and in $[Ag(P\{OEt\}_3)_n]X$, J(AgP) falls from 794 to 535 to 402 to 342 as n increases from 1 to 4. This might easily be explained by changes in the hybridization of the central atom; the amount of s-character in the bond to any given ligand decreases as the number of ligands increases. Although strictly limited to first-row elements, the hybridization model nonetheless provides a useful rule of thumb in considering the magnitude of coupling constants.

Me₃C, H₂C, Me₃C, PMe₃, W≡CCMe₃, PMe₃, H

Figure 3.12 The unusual complex $(PMe_3)_2W(\equiv CCMe_3)(=CHCMe_3)$ $(-CH_2CMe_3)$ contains sp, sp^2, and sp^3 ligating carbon atoms. The tungsten–carbon coupling constant increases with s-character at carbon.

Source: From data in D. N. Clark, et al. (1978), J. Am. Chem. Soc., **100**, 6774.

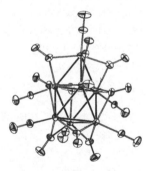

	av J(RhC)/Hz
CO	81
δ^2-CO	40
δ^3-CO	24

Figure 3.13 $Rh_7(CO)_{16}^{3-}$ contains terminal (sp carbon), edge bridging (sp^2 carbon) and face-bridging carbonyls (sp^3 carbon). The magnitude of the rhodium–carbon coupling constant falls as the s-character at the carbon decreases.

Source: Adapted from V. G. Albano et al. (1988), J. Chem. Soc. Dalton Trans., 1103. Reproduced by permission of The Royal Society of Chemistry. Copyright © 1988.

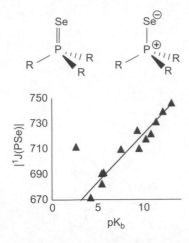

Figure 3.14 Basicity of a selection of phosphane ligands, PR$_3$, (R = alkyl or 4–R'C$_6$H$_4$) vs coupling constant and valence isomers of R$_3$PSe

Source: From data in U. Beckmann et al.(2011), *Phosphorus, Sulfur, and Silicon and the Related Elements,* **186**, 2061–70.

H$^-$ >CH$_3^-$ ~acyl$^-$ >P(OPh)$_3$ ~

PPh$_3$ >CN$^-$ >CO> AsPh$_3$ >

NO$_2^-$ ~SPh$^-$ >SbPh$_3$ >SCN$^-$ >

C$_5$H$_5$N>NCO$^-$ ~>O$_2$CCH$_3^-$ >

Cl$^-$ >CH$_3$CN>ONO$_2^-$

and

C$_2$H$_5^-$ >C$_6$H$_5^-$ >CH$_2$Ph$^-$ ~CH$_3^-$ >

CF$_3^-$ >CH$_2$COCH$_3^-$ >

CH$_2$NO$_2^-$ >CN$^-$ >

CH(COCH$_3$)$_2^-$ >CO.

The metal to ligand coupling constant in complexes is strongly influenced by the ligand in the *trans* position. The *trans*-influence series above for square planar Pt (II) complexes has been derived from $^1J(^{195}$Pt^{31}P) coupling constants.

From data in T. G. Appleton and M. A. Bennett (1978), *Inorg. Chem.,* **3**, 738.

Factors influencing the coupling constant–electronegativity

Coupling constants increase with increasing electronegativity of the substituents in line with a (presumed) increase in *s*-character in the bond. For example, the selenium–phosphorus coupling in organophosphorus selenides, R$_3$PSe, increases as the electronegativity of the substituents on phosphorus increases, Me < Ph < OMe (Table 3.5).

Note also the general trend of increasing downfield shift of the phosphorus as the electronegativity/electron withdrawing power of the substituents increases (see section 3.2, 'Factors influencing the chemical shift—electronegativity, charge, and oxidation state (Table 3.1)). The reason for the reverse trend observed for the selenium chemical shift is unclear and has been attributed to an increase in electron density on selenium as phosphorus substituent electronegativity increases (Figure 3.14).

Similar effects are observed in couplings involving transition elements. For example, ^1J(PtP) in *cis*-[PtCl$_2$(PBu$_3$)$_2$] is 3508 Hz and in *cis*-[PtCl$_2${P(OEt)$_3$}$_2$] is 5698 Hz. Interestingly, the ratio of the phosphine and phosphite coupling constants is fairly independent of the nature of the acceptor; being 1.62 in these platinum complexes and 1.53 in the borane complexes Me$_3$P-BH$_3$, $^1J(^{11}$BP) 64 Hz, and (MeO)$_3$P-BH$_3$, 97 Hz indicating that the ratio of the phosphine-acceptor to phosphite-acceptor coupling constants is determined by the phosphorus *s*-character in the valence bond.

Factors influencing the coupling constant–*trans*-influence

The magnitude of couplings between the ligand and acceptor is strongly influenced by the other ligands in the complex. This effect is particularly evident in square planar complexes of the noble metals. π-acid ligands in the *trans* position reduce the magnitude of the coupling between the metal and the ligand. Thus, in the series of complexes shown in Figure 3.15, ^1J(RhP) falls from over 200 Hz in the *trans*-chloro complex (π-donor) through 140 Hz in the *trans*-phosphine complex to only 96 Hz in the *trans*-carbonyl complex (strong π-acceptor).

Table 3.5 NMR parameters for some organophosphorus selenides

Compound	$^1J_{SeP}$/Hz	δ_{Se}/ppm	δ_P/ppm
Me$_3$PSe	−684	−234.9	8.0
Me$_2$PhPSe	−710	−271.7	15.1
MePh$_2$PSe	−725	−277.0	22.3
(MeO)Me$_2$PSe	−768	−218.6	90.2
(MeO)Ph$_2$PSe	−810	−276.3	86.8
(MeO)$_2$MePSe	−861	−268.9	102.3
(MeO)$_2$PhPSe	−876	−320.4	97.5
(MeO)$_3$PSe	−963	−396.1	77.5

Source: From data in W. McFarlane et al. (1973), *J. Chem. Soc. Dalton Trans,* 2162.

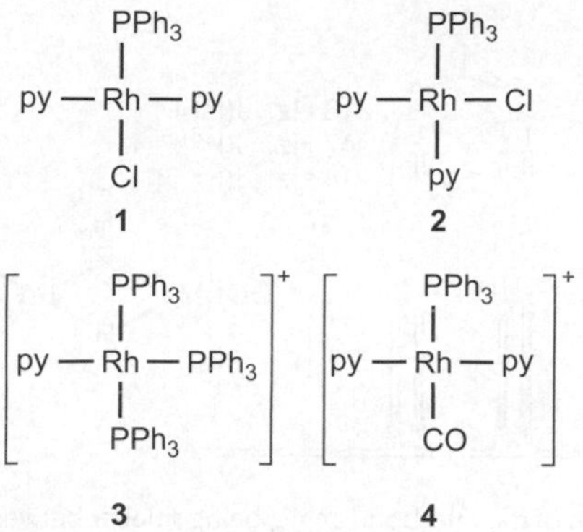

```
        PPh₃                      PPh₃
         |                         |
  py — Rh — py            py — Rh — Cl
         |                         |
        Cl                        py
         1                         2
```

	trans-L	J(RhP)
1	Cl	208
2	py	162
3	py	146
3	PPh₃	161
4	CO	96

```
 [      PPh₃      ]⁺   [      PPh₃      ]⁺
 |      |          |   |      |          |
 [ py — Rh — PPh₃ ]   [ py — Rh — py   ]
 |      |          |   |      |          |
 [     PPh₃       ]   [      CO        ]

        3                    4
```

py = pyridine

Figure 3.15 The *trans*-ligand influences the coupling constant; π-acid and hydride trans-ligands result in smaller couplings; π-donors in the *trans* position give larger values of the coupling constant

Source: From data in B. T. Heaton, et al. (1994), *J. Chem. Soc. Dalton Trans.*, 2875.

However, the explanation for this effect cannot simply be the π interaction with the metal, since hydride ligands also have a high *trans*-influence. In fact, the explanation, once again, is the degree of *s*-character in the bond and the *s*-electron density at the nucleus, which is least for π-acid ligands and for hydride.

Factors influencing the coupling constant–interbond angles

The *s*-character in the bonds between the coupling nuclei is expected to vary with bond angle; simplistically a bond angle of 90° can be obtained using pure *p* orbitals; as the bond angle increases, increasing *s*-character must be added to the bonding hybrid orbital 109° (*sp³*), 120° (*sp²*), 180° (*sp*). We would expect, therefore, that the magnitude of a coupling constant should correlate with the interbond angle(s) between the coupling nuclei, i.e. for two bond couplings, *trans* couplings will normally be larger than *cis* couplings. Thus, for example the coupling between the bridging phosphido ligand and the *trans* phosphine in the palladium hydride complex shown in Figure 3.16 is *ca.* 213 Hz, whereas the coupling to the *cis* phosphine is *ca.* 0 Hz. Similar trends are seen in the phosphorus–proton couplings being 74 Hz when the phosphine ligand is *trans* but only 14–21 Hz when the phosphorus ligand is *cis* (Figure 3.16). With experience, typical values for *cis* and *trans* couplings for any given pair of nuclei can be learnt and become useful in assigning the geometry of new complexes.

For three-bond couplings, not only are the interbond angles important but also the dihedral angle, ϕ, between the planes containing the coupling nuclei and the bonds between them; the dihedral angle dependence of three-bond couplings is expressed in the *Karplus equation* (equation 3.3), which predicts that ^3J will normally be greatest when ϕ is close to 0° or 180°.

In many examples, such as three-bond couplings in transition metal cluster compounds, the effects of both geometrical factors above, *cis–trans* orientation, and dihedral angle on the magnitude of the coupling constant need

Eq. 3.3 The Karplus equation describes empirically the dependence of the coupling constant on the dihedral angle. B is usually negative and $|C| > |B|$, thus ^3J is usually greatest when the coupling partners are *trans*, $\phi = 180°$.

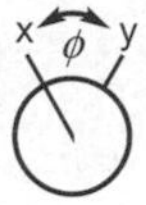

$$^3J = C\cos 2\varphi + B\cos\varphi + A$$

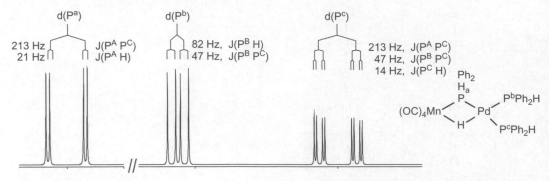

Figure 3.16 The size of the coupling constant is a good indication of the interbond angle, being smaller between similar groups in a *cis* orientation than when the coupling nuclei are mutually *trans*. Here the ^{31}P NMR spectrum of a palladium hydride complex is shown. The larger couplings are between *trans* groups.

Source: From data in P. Braunstein et al. (1992), *Inorg. Chem.*, **31**, 411.

to be considered. Consider $^3J(PC)$ in the fragment P–Rh(X)–Rh(Y)–C(Z) of $Rh_6(CO)_{15}(PPh_3)$ (Figures 1.9 and 3.17).

The terminal carbonyls of the cluster can be divided into six groups; C^4 is clearly unique, being the only terminal carbonyl ligand on Rh^A, and gives the doublet of doublets (coupling to ^{103}Rh and ^{31}P) at low field. Rh^B lies under the triphenylphosphine ligand, whilst Rh^C lies on the opposite side of the cluster; thus the carbonyls on Rh^B must be different from those on Rh^C. The ligands C^7 and C^8 differ, lying respectively above and below the equatorial plane of the cluster. Similar considerations differentiate C^5 from C^6. The two ligands C^9 on Rh^D (not shown in Figure 3.15) are related by symmetry.

$^3J(PC)$ is expected to reach a maximum when there is a *trans-trans* configuration across the intervening Rh–Rh bond and when φ, the dihedral angle between the Rh–C and Rh–P bonds, is close to 0° or 180°. Conversely, $^3J(PC)$ will be at a minimum when there is a *cis-cis* configuration with φ close to 90°. There are four possible configurations of the P–Rh(X) –Rh(Y)–C(Z) fragment (Table 3.6).

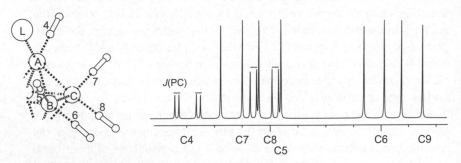

Figure 3.17 Expansion of part of the structure of $Rh_6(CO)_{15}(PPh_3)$. The magnitude of $^3J(PC)$ between the phosphorus ligand, L, and carbonyls C(5)–C(8) depends on the dihedral angle and the *cis/trans* orientation of the intervening bonds. All resonances show coupling to ^{103}Rh and C(4) shows $^2J(PC)$; however, only C(8) is in a *trans-trans* orientation to the phosphine ligand. $^3J(PC)$ coupling is only resolved for this pair of nuclei.

Table 3.6 The relationship between interbond angles and dihedral angles for $^3J(PC)$ in $Rh_6(CO)_{15}(PPh_3)$

Configuration	Interbond angles	Dihedral angle	Predicted $^3J_{PC}$	Found $^3J_{PC}$ (Hz)
L-Rh^A-Rh^B-C^5	cis–cis	10	Small	0
L-Rh^A-Rh^B-C^6	cis–trans	97	Small	0
L-Rh^A-Rh^C-C^7	trans–cis	97	Small	0
L-Rh^A-Rh^C-C^8	trans–trans	30	Large	20

Source: From data in E. V. Grachova et al. (2001), *J. Chem. Soc. Dalton Trans*, 3303–11.

Only in the last do the geometric factors reinforce each other, so we would expect to see the maximum value of $^3J(PC)$ between the phosphine ligand and C^8. In fact, three-bond coupling is only seen for C^8, $^3J(PC) = 20$ Hz, allowing ready assignment of the higher field doublet of doublets to this ligand. All other resonances can be assigned using the rhodium–carbon couplings.

Factors influencing the coupling constant–lone pairs

The presence of lone pairs usually makes the coupling constant more negative; however, when the lone pair is used in co-ordination the contribution from the lone pair becomes positive. We would therefore expect the *magnitude* of the coupling constant to decrease on co-ordination. For example, the $^1J(PSe)$ in free Ph$_3$PSe is -655 Hz, becoming −585 Hz on co-ordination to cadmium. Similarly, the magnitude of $^1J(PSe)$ in Bu$_3$PSe falls on co-ordination to HgX$_2$. The decrease in $|^1J(PSe)|$ is most pronounced for the most Lewis acidic mercury halide as expected; the lone pair becomes more intimately involved in bonding to the metal (Table 3.7).

Similarly, in the phosphazene complex AuCl{μ-Ph$_2$PNPPh$_2$PPh$_2$NPPh$_2$}AuCl the magnitude of the phosphorus–phosphorus coupling constant falls from *ca.* 85 Hz in the free ligand to only 13 Hz in the complex (Figure 3.18), whereas the P–N–P interbond angle is virtually unchanged falling from 131° to 129°, indicating the change in $^2J(PP)$ is a result of complexation of the lone pair by the metal.

Table 3.7 The magnitude of $^1J(SeP)$ decreases on co-ordination due to involvement of the Se lone pair in bonding to the metal

Compound	$|^1J(SeP)|$ /Hz
(n-Bu)$_3$PSe	693
{(n-Bu)$_3$PSe}$_2$HgCl$_2$	551
{(n-Bu)$_3$PSe}$_2$HgBr$_2$	560
{(n-Bu)$_3$PSe}$_2$HgI$_2$	587

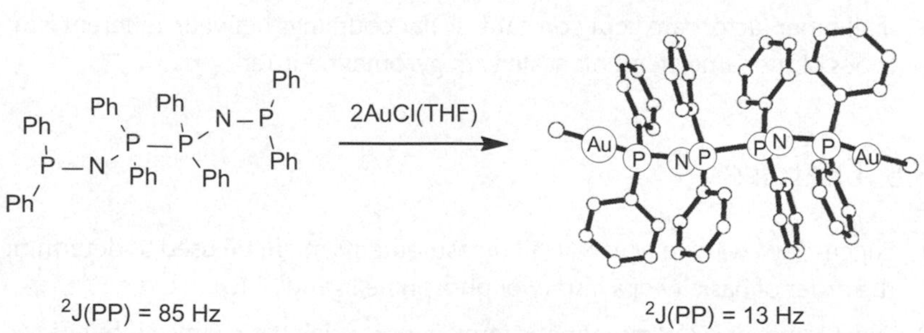

$^2J(PP) = 85$ Hz $^2J(PP) = 13$ Hz

Figure 3.18 Co-ordination of the phosphorus lone pair in phosphazene ligands results in a reduction in the magnitude of the PP coupling constant

Sources: Adapted from J. D. Woollins et al. (1996), *J. Chem. Soc., Chem. Commun.*, 2095, reproduced by permission of The Royal Society of Chemistry Copyright © 1996 and data in P. Braunstein et al. (1995), *J. Chem. Soc., Chem. Commun.*, 37.

$^1J(PtP) = 3740$ Hz

$^1J(PtP) = 2411$ Hz

$^1J(PtPBu_3) = 3159$ Hz
$^1J(PtP(OPh)_3) = 6304$ Hz

$^1J(PtPBu_3) = 1951$ Hz
$^1J(PtP(OPh)_3) = 4060$ Hz

Figure 3.19 The coupling between the metal and the ligand decreases as the oxidation state of the metal increases

Source: From data in D. E. Gerlach et al. (1971), *J. Amer. Chem. Soc.*, **93**, 3543 and F. H. Allen et al. (1971), *J. Chem. Soc. A.*, 2054.

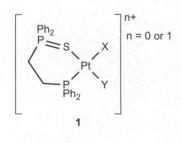

1

The dihedral angle between two lone pairs in for example diphosphides, R_2PPR_2, also affects the magnitude of the PP coupling, which is greatest when the lone pairs are mutually *trans*.

Factors influencing the coupling constant—oxidation state

Empirically it is found that scalar coupling reduces as the oxidation state of the coupled nucleus increases. For example, the coupling between platinum and phosphorus falls in the ranges 1050–2075 Hz for $Pt^{IV}(PR_3)$, rising to 1675–5850 Hz in platinum(II) complexes, reaching 2825–9150(!) Hz for $Pt^0(PR_3)$ (Figure 3.19).

These trends hold notwithstanding changes in geometry and co-ordination number, although the nature of the donor (phosphine versus phosphite) and changes in the *trans* ligand do affect the size of the coupling. This trend might have been expected, since coupling is transmitted through the polarization of the electrons in the bonds. As the oxidation state is increased this becomes more difficult to achieve, reducing the magnitude of the coupling.

3.4 Summary

- Shielding is influenced by the fields generated by circulation of ground state electrons around the nucleus (diamagnetic term) and by electrons mixed in from excited states (the paramagnetic term). The diamagnetic contribution is usually small and dominates for light elements, resulting in (usually) small chemical shift ranges for these elements. The paramagnetic term dominates for heavier elements and is often large, resulting in large chemical shift ranges for these elements.

- The chemical shift is influenced by many, often competing factors including oxidation state, electronegativity, coordination number, and bonding to other atoms.

- Scalar coupling constants, J, are influenced by similar factors and by geometric factors, particularly interbond and dihedral angles.

- If all other factors are kept constant, scalar couplings between different isotopes of the same elements scale with gyromagnetic ratio, γ.

3.5 Exercises

1. Suggest two ways in which NMR measurements might be used to determine the order of basicity of a library of phosphine ligands, PR_3.

2. What is meant by 'paramagnetic term' in discussing the chemical shift of transition metals?

3. A series of Pt complexes **1** has been prepared where one of **X** and **Y** is a methyl group and the other is one of the following ligands: acetone, acyl,

SnCl$_2$ or CO. $^1J(^{195}Pt^{31}P)$ in the complexes varies from 3367 to 3995 to 4036 to 5003. By considering the effect of *trans*-influence, decide whether **X** or **Y** is the methyl group. Pair each ligand with the appropriate $^1J(^{195}Pt^{31}P)$.

4. Sketch stick diagrams to illustrate the couplings present in the ^{1}H NMR spectrum of each isotopologue of **2**. $^1J(^{109}Ag^1H)$ in the dinuclear silver compound **2** is 134 Hz. Use the data in Table 1.2 to calculate $^1J(^{107}Ag^1H)$. From data in B. K. Tate et al. (2013), *Chem. Sci.*, **4**, 3068.

Experimental methods: pulses, the vector model, and relaxation

4.1 Introduction

So far we have not considered how an NMR spectrum is recorded or how we can tailor the spectroscopic experiment to maximize the information we are interested in while suppressing unwanted complexity. Modern, multiple-pulse NMR methods can do this. Before we venture into this territory, we need to consider how we can describe what is going on in our tailored experiment. We also need to look briefly at relaxation, which is not only an important factor to consider when recording spectra but can also be used to obtain chemical information about our sample.

4.2 Experimental methods

Before exploring multiple-pulse methods, we first need a basic understanding of how the NMR signal is created and NMR spectra acquired. Traditionally this was done using the continuous wave (CW) method. Modern spectrometers use the pulse Fourier transform (FT) method.

Continuous wave NMR spectroscopy

In the CW method either the magnetic field or the irradiation frequency is adjusted to bring each nucleus in turn on to resonance. The net absorption of energy at resonance is detected and a spectrum produced. This method is analogous to conventional IR or UV spectroscopy. To achieve good resolution of closely spaced resonances the sweep rate must be slow, typically taking five to twenty minutes; signal averaging—recording the spectrum many times to improve the signal-to-noise ratio is impractical. The low sensitivity of NMR restricts the CW method to samples that give strong NMR signals, i.e. to concentrated samples of nuclei with high gyromagnetic ratios, in effect protons and fluorine.

The Fourier transform method

If all the resonances in a sample could be excited simultaneously and the resultant signal analysed quickly, i.e. in a few seconds, it would be possible to measure the NMR spectrum many times, add the results together to improve the signal-to-noise ratio, and so obtain NMR spectra of dilute solutions and for nuclei with low receptivity. This is what the FT method achieves.

Consider the following analogy: to discover the resonance frequencies of a bell, we could borrow a sine wave generator and patiently try every frequency, listening for the bell's resonances (Figure 4.1). This is the CW method: the bell's resonances are measured one by one in the frequency domain, a very time-consuming process. Alternatively, we could borrow a frequency analyser, hit the bell to excite all the bell's resonances, and analyse the ringing of the bell (Figure 4.2); *all* the bell's resonances are measured concurrently in the time domain; all frequencies are tried simultaneously *and* we need only analyse the ring for a few seconds to differentiate frequencies less than a hertz apart—the method is very time-efficient. We can convert from the time domain to the more familiar frequency domain by a Fourier transformation, hence the method is dubbed the FT method. Because we let the bell ring *freely* after striking it, because we had to strike the bell to *induce* it to ring, and because the ringing of the bell dies away or *decays*, the signal we analyse is called the **free induction decay** or **FID**. Using the FT method, it is quite feasible to record thousands of FIDs, add them together, and then Fourier transform the result to obtain the NMR spectrum of even insensitive nuclei. In NMR, we use a pulse of radiation to excite the nuclear spins; using more than one pulse allows us to manipulate the spins to edit the information in the measured spectrum—multiple-pulse NMR—greatly increasing the analytical power of the technique.

Figure 4.1 The continuous wave method is very slow, each frequency must be tried separately to find the resonance frequency.

Figure 4.2 The FT method excites all resonances simultaneously and is therefore very quick. The resonances can be excited and measured many times to improve the signal-to-noise ratio.

4.3 The vector model and the rotating frame

To understand what is happening in a pulse NMR experiment, it is instructive to consider how the magnetization of the sample is affected by the pulse. A simple picture that gives a feel for this is the **vector model**. More advanced descriptions, such as the spin density matrix, are beyond the scope of this book.

Magnetic moment of a collection of spins

Up to now we have considered only a single molecule which aligns its nuclear magnetic moment at certain fixed angles with or against the magnet field. In a real sample, of course, there is more than one molecule and the molecules are not obligingly lined up with the applied field; in a solid the molecules are held in fixed positions in the lattice, whilst in solution Brownian motion means that the molecules are tumbling randomly. Let's look at the nuclear magnetic moments of all the molecules in a liquid sample. At first sight the moments appear to be randomly oriented in all directions (Figure 4.3); however, if we look more

Figure 4.3 The nuclear magnetic moments in a liquid sample appear randomly orientated but there is an excess of magnetization aligned with the field.

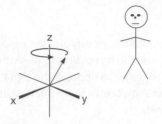

Figure 4.4 The nuclear magnetic moments of the spins will precess about the applied field

Figure 4.5 In the rotating frame, the z′-axis is placed along the magnetization vector and precesses with it at the Larmor frequency—the magnetization appears stationary along z′.

The picture in the laboratory frame is much more complicated. At equilibrium the nuclear magnetization is precessing about the z-axis. Application of the excitation pulse causes a second precession of the magnetization about an axis at the same angle as was the equilibrium magnetization which is still precessing about z. The motion of the magnetization vector is hard to describe(!) and harder still to visualize.

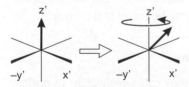

Figure 4.6 In the rotating frame excitation of the spins displaces the nuclear magnetic moment away from the z-axis. Imagine grasping the axis about which the rotation is taking place with your right hand, thumb along the positive axis; your fingers curl in the sense of a positive rotation.

closely, we find that at any given time there is an excess of nuclear magnetization aligned with the applied field. There will therefore be a macroscopic nuclear magnetic moment associated with the sample. Like the individual nuclear moments, this macroscopic moment can be represented by a vector. Since the individual moments of which the macroscopic moment is comprised are required by quantization to be at an angle to the applied field, the macroscopic moment is also at an angle to the field and so it must precess about the field at the Larmor frequency.

Precession of the nuclear spin

Classically, a spinning body—say a child's spinning top—in a static field—say the Earth's gravity—will precess, i.e. it will rotate on the surface of a cone whose axis is in the field direction. Nuclear magnetic moments precess about an applied magnetic field for the same reason (Figure 4.4). The rate of precession is known as the *Larmor frequency* and can be calculated classically or quantum mechanically. It turns out that the Larmor frequency is also the resonance frequency (Chapter 1, section 1.4).

The rotating frame

To simplify the physical picture of what happens to the macroscopic magnetization during the NMR experiment we move from the static laboratory frame of Figures 4.3 and 4.4, to a frame of reference rotating at the Larmor frequency and with the z′-axis inclined to the z-axis at the same angle as the equilibrium magnetization. Basically, this means we will sit on the end of the vector describing the magnetization and look at the NMR experiment from there, ignoring the world as it flashes around us (Figure 4.5). This makes the Larmor precession appear to 'vanish'. Many analogies have been given to illustrate the simplification of the physical picture achieved by moving to the rotating frame, for example describing the motion of the Moon as viewed from the Sun (the laboratory frame) or the Earth (the rotating frame). Viewed from the Sun, the Moon appears to follow a complex path as it orbits the Earth; viewed from the Earth, it simply goes round and round.

The vector model of NMR considers what happens to the bulk magnetization vector during the NMR experiment, as viewed in the rotating frame. For simplicity, from here on we will refer to the axes in the rotating frame as x, y, and z, dropping the "′" designator.

The excitation pulse

In the FT method, the NMR experiment begins by applying a pulse of radiation (energy) of the correct frequency to induce transitions between the upper and lower nuclear spin energy levels. The transitions induced in the individual nuclear spins change the macroscopic magnetization of the sample and this change appears as a displacement of the magnetization from the z-axis (Figure 4.6).

Different texts/authors choose different conventions so an *x*-pulse might be shown rotating the magnetization towards −*y* or +*y*, depending on the author. This is not really important, so long as a consistent convention is used; here we follow the convention that a +*x* pulse will rotate the magnetization toward the −*y* axis, and a +*y* pulse to the +*x*-axis (Figure 4.6). The intensity of the excitation pulse will determine how much the magnetization vector is tipped away from the *z*-axis, the tip angle. The motion of the magnetization vector after the excitation pulse is easily visualized; it will precess about the *z*-axis at a rate that depends on the offset of the chemical shift from the Larmor frequency. The NMR signal that we observe is the projection of this precession onto the *x*–*y* plane.

We can identify two important pulse lengths: a 90° pulse will bring the magnetization vector into the *x*–*y* plane and corresponds to the maximum intensity of the NMR signal; whilst a 180° pulse aligns the magnetization along the minus *z*-axis, the magnetization is said to be inverted and there is no observable magnetization (the projection on the *x*–*y* plane is zero).

Vector model of chemical shifts

The effect of the chemical shift is to split the macroscopic moment into several components—replace the vector representing the macroscopic moment with a packet of vectors to represent the different chemical shifts. Each chemical shift vector will have a slightly different rate of precession, corresponding to the different resonance frequencies of the chemically different nuclei present. This is easily visualized in the rotating frame; at equilibrium the magnetization is aligned along the *z*-axis. The excitation pulse displaces the magnetization from the *z*-axis so it starts to precess as before (Figure 4.6). However, the presence of different chemical shifts, i.e. of nuclei with differing resonance frequencies, results in our packet of chemical shift vectors fanning out into as many components as there are chemical shifts (Figure 4.7).

Vector model of scalar couplings

Scalar couplings have a similar effect on the magnetization vector as chemical shifts: this time, the magnetization splits into as many components as there are lines in the multiplet, the downfield lines precessing faster than the upfield lines. There is an important difference, though. While the different chemical shifts each 'do their own thing', in the case of scalar coupling, the precession rates are not independent; the difference in the precession rates will be the coupling constant. Figure 4.8 illustrates this for a doublet; the two magnetization vectors precess at the same rate, $J/2$, but in opposite directions; after time τ each of the two vectors has precessed by $\pi J\tau$ radians. Thus, at time $1/2J$ after an *x*-pulse, the magnetization vectors of the two branches of the doublet have each precessed by $\pi/2$ radians and will lie over the +/−*x*-axis (but pointing in opposite directions), will line up over *y* after time $1/J$, and finally come together again over −*y* at time $2/J$, when each vector will have precessed through 360° but in opposite directions.

Excitation, saturation, and relaxation

Excitation changes the populations of the spin energy levels away from the equilibrium values. If the populations of the energy levels in an NMR transition are equalized, the transition is said to be saturated and the NMR signal and *J* couplings to the saturated resonance disappear. Relaxation is the process(es) by which an excited or saturated system loses energy/returns to equilibrium.

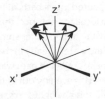

Figure 4.7 Chemical shifts cause the magnetization to fan out. Here the magnetization splits into four different chemical shifts. The four chemical shift vectors precess at independent rates, causing the magnetization to fan out.

Figure 4.8 Scalar coupling splits the magnetization. Here we see the two components of a doublet. The two components precess at the same rate, $J/2$, but in opposite directions.

4.4 Relaxation

In CW NMR the spectrum is recorded in absorption mode; in FT NMR we first excite the sample with a pulse of radiation and then look at the emission signal. For multiple-pulse and for quantitative experiments, it is important that we start each experiment from equilibrium; if we do not, then our experiment may give no spectrum at all or a completely wrong result.

Relaxation is the process by which excited nuclear spins lose energy and drop back to the ground state. Nuclear spins can get from the upper to the lower level in several ways. Up until now we have considered only radiative transitions, the 'direct route', i.e. those that obey the NMR selection rule. However, it is also possible for a nucleus to relax by giving up its energy non-radiatively either to other nuclear spins—**spin–spin relaxation**—or to its surroundings—**spin–lattice relaxation**. In spin–spin relaxation energy is exchanged between nuclear spins but the total amount of excited spin in the system does not change, whereas spin–lattice relaxation re-establishes Boltzmann equilibrium. The difference between these relaxation processes can be likened to what happens after a bucket of boiling water is poured into a bath (Figure 4.9). The heat from the added water first causes a 'hot spot' before 'spreading out' through the water already in the bath. This evens out the temperature of the bath water but does not remove heat from the bath—an analogy for spin–spin relaxation. The bath also slowly cools down, back to its starting temperature, as heat is lost to the bathroom—this is spin–lattice relaxation. Both processes are normally exponential in nature and are referred to by their time constants, T_2 and T_1 respectively. The reciprocal of the time constant is the rate constant for the relaxation. It is possible for magnetization to be exchanged between spins faster than Boltzmann equilibrium is re-established, but clearly the reverse is impossible; hence $T_2 \leq T_1$.

T_1 relaxation—how long the system takes to re-establish equilibrium—is important; it determines how frequently we can repeat a measurement to improve the signal to noise—for NMR this can be less than a millisecond or more than an hour (!) depending on the nucleus and its environment. Typically, we need to wait from 1 to 5 T_1s between experiments to allow sufficient time for T_1 relaxation. More than 5 T_1s are needed to re-establish Boltzmann equilibrium so, in multi-pulse experiments or if we wish to quantify an NMR experiment, it may be necessary to wait $10 \times T_1$ before repeating a measurement. T_2 relaxation tells us about the lifetime of the excited state and hence, *via* the uncertainty principle, the natural linewidth of the NMR line. Relaxation can also tell us about the chemistry going on in the system, so measurement of relaxation rates is an important chemical tool.

Relaxation mechanisms

Relaxation occurs when the nuclear spin interacts with a magnetic field oscillating at or near its resonance (Larmor) frequency. The sources of such fields include

Figure 4.9 After a bucket of boiling water is poured into a bath, at first there is a localized hot spot, then, as heat from the added water spreads through the water already in the bath the temperature of the bath water evens out. This evening out of the temperature is analogous to spin–spin relaxation. The bath also slowly cools back to its original temperature as heat is lost to the surroundings—this is analogous to spin–lattice relaxation.

Eq. 4.1 If relaxation occurs by more than one mechanism the reciprocal of the observed time constant, T_{obs}, will equal the sum of the reciprocals of the individual time constants.

$$\frac{1}{T_{1obs}} = \frac{1}{T_{1dd}} + \frac{1}{T_{1csa}} + etc$$

the spins of other nuclei, unpaired electron spins, and nuclear quadrupoles. The first two are examples of dipole–dipole relaxation, the last is called **quadrupolar relaxation**. Relaxation by, for example, **chemical shift anisotropy** and **chemical exchange** is also possible. If relaxation occurs by more than one mechanism, the observed relaxation rate will be the sum of the rates due to each mechanism, i.e. the reciprocal of the observed time constant, T_{obs} will equal the sum of the reciprocals of the individual time constants (equation 4.1). When measuring relaxation, it is important to exclude unwanted processes, for example relaxation by the electron spins of dissolved oxygen which has a paramagnetic ground state.

Dipole–dipole relaxation

Dipole-dipole relaxation occurs when the sample contains other magnetic dipoles that can interact with the nuclear magnetic moment. These can be the magnetic moments of other nuclei or of unpaired electrons. The effectiveness of the relaxation depends on the square of the gyromagnetic ratios of the two spins, so high γ nuclei will give faster relaxation than low γ nuclei. The gyromagnetic ratio of the electron is about 1000 times that of the proton, thus when unpaired electrons are present, for example in some transition-metal ions, relaxation will be dominated by dipole–dipole relaxation from these electrons.

To cause relaxation the dipoles must set up a magnetic field that oscillates at or near the NMR resonance frequency. This oscillation is produced by, for example, the Brownian motion of the molecules. It is usual to think of this motion in terms of a correlation time, τ_c. This rather nebulous concept is the time it takes for the magnetic moment to re-orientate in space the smallest significant amount relevant to the relaxation process! Clearly different molecules will be able do this more or less easily depending on factors such as size, shape, and the viscosity of the medium, so relaxation can tell us about such properties. τ_c will also be temperature dependent, so the dipole–dipole relaxation rate will vary with temperature and we can extract an activation energy using the Arrhenius relationship. Spin–lattice relaxation is less effective, corresponding to a long T_1, both at short correlation times, the **fast-motion regime**, and at long correlation times, the **slow-motion regime**. The T_1 minimum (most efficient relaxation) occurs when $\tau_c = (\omega_0 \sqrt{2})^{-1}$. At short correlation times, T_2 equals T_1 but in the slow-motion regime, T_2 continues to fall until a limiting value is reached (Figure 4.10). Note the intimate relationship between relaxation, mobility, and frequency, which we shall return to shortly.

Quadrupolar relaxation

The importance of quadrupolar relaxation depends on the size of the nuclear quadrupole and the electric field gradient at the nucleus. The latter depends on the symmetry of the molecule and is smallest in cubic environments. Where the

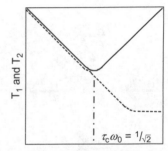

Correlation time increasing
or
temperature decreasing

Figure 4.10 The relaxation rate varies with temperature due to the temperature effect on Brownian motion and hence on τ_c. Notice that T_2 (dotted line) $\leq T_1$.

Source: Adapted with permission from N. Blombergen et al., Physical Review, 1948, 73, 679. ©1948 American Physical Society.

quadrupole is large and/or the environment far from cubic symmetry, quadru-polar relaxation is very efficient and has the effect of making some quadrupolar nuclei essentially NMR invisible. The very short lifetime of the NMR states results in severe Heisenberg broadening (equation 1.5), giving NMR lines many thousands of hertz wide—the NMR spectrum becomes a broad hump barely visible above the baseline (Table 1.3, section 1.8). If quadrupolar relaxation is very efficient it can also effectively decouple other spins from the quadrupolar nucleus: the quadrupolar nuclear spin flicks between its different possible orientations too rapidly for the observing nucleus to sense the different orientations. In these cases, the only likely effect of the presence of a quadrupolar nucleus in the sample will be to enhance the relaxation of neighbouring spins.

Effect of relaxation on the NMR spectrum

Unless care is taken, relaxation effects can be a nuisance since incomplete relaxation affects the measured intensity of each resonance differently. For example, the widely differing, and slow, relaxation rates of primary, secondary, tertiary, and quaternary carbons prohibit the use of line integrals in $^{13}C\{^1H\}$ NMR as a guide to the number of each type of carbon atom in a molecule unless unfeasibly long inter-scan delays are used to allow equilibrium to re-establish. This effect can also be seen in ^1H NMR spectra of organic compounds. Aryl protons tend to relax more slowly than alkyl protons; if too short a delay is used between scans the integrals of the aryl protons will be less than expected compared to those of alkyl protons.

Nuclear Overhauser effect—the nOe

Another reason integration is not used in $^{13}C\{^1H\}$ NMR spectroscopy is the nuclear Overhauser effect (Figure 4.11). This results from cross-relaxation caused by through-space dipole–dipole interactions of the ^1H and ^{13}C spins that are close in space to each other. Proton decoupling irradiation (ν_1) equalizes the proton spin populations, thus increasing the populations of the $\beta\alpha$ and $\beta\beta$ levels (dotted lines in Figure 4.11). Dipole–dipole interactions cause ω_2 relaxation directly from the $\beta\beta$ state to the $\alpha\alpha$ state (and from the $\alpha\beta$ to $\beta\alpha$ states). If ω_2 is the dominant relaxation pathway, then every time a proton spin relaxes, a carbon spin relaxes with it (Figure 4.11, bottom). This changes the populations across the ^{13}C transitions, increasing the number of carbon α spins at the expense of carbon β spins, i.e. increasing the population difference across the ^{13}C energy levels which, in turn, increases the intensity of the ^{13}C NMR resonances of *only those carbons that are close to protons*. The intensity of a ^{13}C resonance thus does not depend only on the number of carbon atoms of each type present in the molecule.

Since the nOe is caused by through-space relaxation of one nucleus by its neighbours and decays as $1/r^6$, the more neighbours and the closer together the two nuclei are, the stronger will be the nOe. nOe is thus a powerful technique to determine molecular conformation and is widely used in organic and protein chemistry (see Chapter 5).

nOe can also be observed in inorganic systems. For example, in the PPh$_2$H ligand in the dimanganese compound shown in Figure 4.12, a strong nOe on

Differences in relaxation rate between nuclei at different sites in the molecule distorts resonance intensities. This is one reason why $^{13}C\{^1H\}$ NMR integrals are not a good guide to the number of carbons at each site in a molecule. A second reason is the nuclear Overhauser effect, which increases the intensity of carbons close to protons through cross-relaxation.

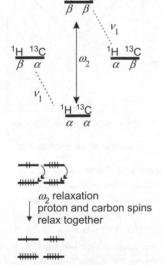

ω_2 relaxation
proton and carbon spins relax together

Figure 4.11 Dipole–dipole interactions result in relaxation directly from the $\beta\beta$ to the $\alpha\alpha$ state; when a proton spin relaxes, it takes a carbon spin with it. This results in an increase in the population difference of the carbon spin energy levels, increasing the intensity of the ^{13}C NMR transition.

phosphorus is observed from the directly bonded proton doubling the intensity of the $^{31}P\{^1H\}$ resonance of this ligand with respect to that of the phosphido bridge and, if not recognized, might mislead the observer into thinking two PPh_2H ligands are present.

Just as in organic and protein structure determination, nOe can be used to determine molecular conformation and the arrangement of molecules or ions within inorganic supramolecular aggregates such as ion pairs. Today, nOes are most frequently determined using the 2-D NMR methods NOESY and HOESY; examples are given in Chapter 5.

Vector model picture of relaxation

In the vector model, the NMR signal arises from the projection of the precessing nuclear magnetic moment on the x–y plane. In some samples this precession can last for hours but clearly the excitation must decay as energy is radiated to give the NMR spectrum, is lost to the lattice, or is exchanged with other spins present in the sample. The NMR signal will vanish if the projection of the moment on the x–y plane vanishes; this corresponds to spin–spin relaxation—the x and y components of the magnetization dephase or loose coherence as energy is exchanged between spins. This happens independently of relaxation of the z component of the magnetization. Relaxation in the z direction re-establishes Boltzmann equilibrium; this is spin–lattice relaxation. A consideration of the geometry of the situation reveals that spin–spin relaxation must be at least as fast as spin–lattice relaxation but that spin–lattice relaxation cannot be faster than spin–spin relaxation (Figure 4.13).

$T_1(rho)$–relaxation in the rotating frame

Now that we have a picture of relaxation using the vector model, we can introduce one more important relaxation parameter—relaxation in the rotating frame, $T_1(rho)$. To understand what $T_1(rho)$ is and why it can be useful it is simplest to

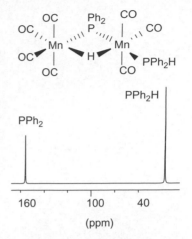

Figure 4.12 The intensity of the $^{31}P\{^1H\}$ NMR resonances in $HMn_2(PPh_2)(CO)_7(PPh_2H)$ is not a good guide to the number of each type of phosphorus ligand due to a strong nOe from the directly bonded proton in PPh_2H

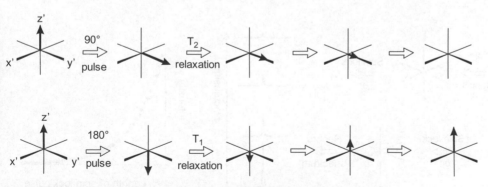

Figure 4.13 Spin–spin relaxation corresponds to the magnetization 'fading away' in the xy plane, shown here after a 90°, $-x$ pulse. When spin–spin relaxation is complete, there is no observable component of the magnetization. Spin–lattice relaxation corresponds to the magnetization 'growing back' along the z-axis, shown here after a 180° pulse. Only when spin–lattice relaxation is complete has the system returned to its initial state, i.e. Boltzmann equilibrium has been re-established.

ignore the—slightly confusing—name and instead look at the vector model of the experiment. In Figure 4.13 we looked at relaxation of the magnetization in the x–y plane (T_2) and return to Boltzmann equilibrium (T_1) after an excitation pulse when the magnetization evolves about the spectrometer field, $\mathbf{B_0}$. In the $T_1(rho)$ experiment we first apply a 90° pulse to tip the magnetization into the x–y plane and then apply a long pulse, called a spin–lock pulse, that generates a constant magnetic field, $\mathbf{B_1}$, orthogonal to the magnetization. The magnetization will start to precess about the $\mathbf{B_1}$ field generated by the spin–lock pulse and we can monitor its relaxation *about $\mathbf{B_1}$* (Figure 4.14). This is useful since relaxation is sensitive to molecular motion at the precession frequency, which in turn depends on the magnetic field (equations 1.1–1.3). In the case of T_1, the relevant magnetic field is $\mathbf{B_0}$, but in the case of $T_1(rho)$ it is $\mathbf{B_1}$. Inspection of Figure 4.14 reveals that $T_1(rho)$ can be measured simply by fitting the intensity of the NMR signal as a function of the length of the spin–lock pulse to an exponential decay (equation 4.2). Since $\mathbf{B_1}$ is several orders of magnitude smaller than $\mathbf{B_0}$, this allows us to probe the molecular motions contributing to relaxation on two different timescales: MHz in the case of T_1, kilohertz for $T_1(rho)$. $T_1(rho)$ is used extensively to study motions in the solid state.

Measurement of T_1, T_2

Inspection of Figure 4.13 also suggests a method for measuring T_1, the spin–lattice relaxation time constant. We first invert the magnetization with a 180° pulse, then wait a delay, τ, to allow some relaxation to occur. A 90° pulse is then used to rotate the magnetization into the x–y plane before measuring the spectrum (Figure 4.15). The intensity of the NMR signal will reflect the amount of spin–lattice relaxation that has taken place. If we vary τ we can determine the intensity as a function of time τ. The time constant, T_1, is found by fitting the observed intensity to equation 4.3.

Eq. 4.2 $T_1(rho)$ and T_2 are found by fitting the measured intensity to a simple exponential decay.

$$M_t = M_0 e^{-\tau/T_2}$$

Eq. 4.3 T_1 is found by fitting the measured intensity to a simple exponential. T_1 can be calculated from the time when the zero-crossing point occurs. $T_1 = \tau/\ln2$ or graphically from the linearized data or by non-linear regression.

$$M_t = M_0(1 - 2e^{-\tau/T_1})$$

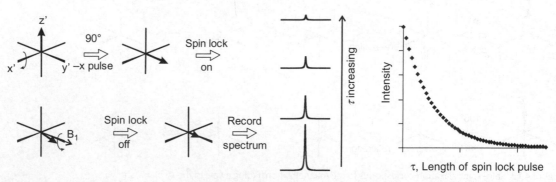

Figure 4.14 $T_1(rho)$ relaxation corresponds to the magnetization 'fading away' about a spin–lock pulse. When relaxation is complete, there is no observable component of the magnetization. $T_1(rho)$ allows study of molecular motion on the kHz timescale, whereas T_1 and T_2 are caused by motions on the MHz timescale.

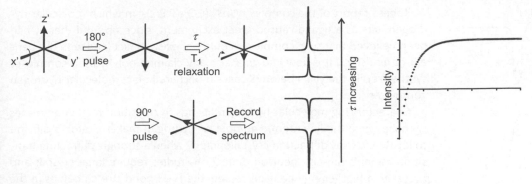

Figure 4.15 A multiple pulse experiment is used to measure T_1. After a 180°, −x pulse a variable delay is inserted to allow the magnetization to 'grow back' part way along the z-axis. A spectrum is recorded after a 90°, −x pulse. The intensity of the spectrum is a function of the relaxation rate (equation 4.3).

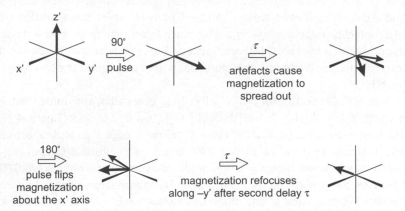

Figure 4.16 A spin echo is used to measure T_2. The intensity of the echo is less due to the spin–spin relaxation. By varying the delay τ the spin–spin relaxation rate can be determined (equation 4.2).

T_2, the spin–spin relaxation time constant, can in theory be measured directly from the linewidth or from the rate of decay of the FID. In practice it is more usual to use a spin echo method that periodically refocuses the magnetization along the y-axis; this removes artefacts such as magnetic field inhomogenity from the measurement (Figure 4.16). The loss in intensity of the resonance then reflects spin–spin relaxation. T_2 is found by fitting the measured intensity to a simple exponential (equation 4.2).

4.5 Application of relaxation measurements to chemical problems

Molecular hydrogen complexes

Relaxation is not merely something we must allow for in recording an NMR spectrum; it can be crucial in characterizing complexes.

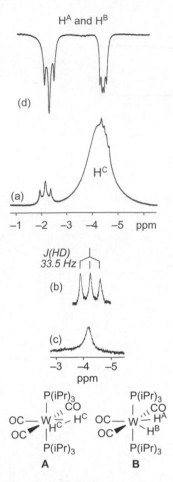

H^A and H^B

H^C

(d)

(a)

−1 −2 −3 −4 −5 ppm

J(HD)
33.5 Hz

(b)

(c)

−3 −4 −5
ppm

P(iPr)₃
| CO
OC — W ⫶ — H^C
OC ◢ | H^C
P(iPr)₃

A

P(iPr)₃
| CO
OC — W ⫶ H^A
OC ◢ | H^B
P(iPr)₃

B

Figure 4.17 The ¹H NMR spectrum (upfield region only) of a mixture of W(CO)₃(PⁱPr₃)₂(H₂), A, and W(CO)₃(PⁱPr₃)₂(H)₂, B. (a) At 194 K the broad resonance of the molecular hydrogen ligand overlaps the resonance of one of the two inequivalent hydrides of the classical complex; (b) coupling to deuterium is seen when HD is used in place of H₂; (d) T₁ edited spectrum showing the resonances of B only.

Source: Reproduced with permission from G. J. Kubas et al. (1986), *J. Amer. Chem Soc.*, **108**, 7000. Copyright © 1986 American Chemical Society. See also G. J. Kubas, et al. (2014), *J. Organomet. Chem.*, **751**, 33–49.

Kubas's report of the complex [W(H₂)(CO)₃(PⁱPr₃)₂] in which molecular hydrogen acts as a ligand caused great excitement, since up until then it had been believed that such complexes could not exist. In fact (and with the benefit of hindsight) it is now known that many claimed polyhydride complexes synthesized in the years before Kubas's report are in fact molecular hydrogen complexes.

Characterizing molecular hydrogen ligands in transition metal complexes is challenging: the low atomic number of hydrogen makes H atoms difficult to locate by X-ray diffraction crystallography when a strongly diffracting transition metal is present; neutron diffraction studies require large crystals and access to a high-energy neutron source; the H–H bond stretch comes in the range 3100–2400 cm⁻¹—a region in which C–H bonds also absorb. Both molecular hydrogen and classical hydride ligands resonate in the same region of the ¹H NMR spectrum—molecular hydrogen ligands between 0 and −15 ppm, and classical hydrides between −2 and −50 ppm. However, the relaxation rates of a molecular hydrogen ligand will be much faster than those of the classical hydride due to the short distance between the H nuclear spin dipoles in the former. Relaxation rate can, therefore, be used to distinguish the two types of ligand.

The ¹H NMR spectrum of '[W(CO)₃(PⁱPr₃)₂H₂]' reveals that a mixture of tautomers [W(CO)₃(PⁱPr₃)₂(H₂)], **A**, and [W(CO)₃(PⁱPr₃)₂(H)₂], **B**, is present (Figure 4.17a). The inequivalent hydrides of the classical tautomer appear as a pseudo-triplet and a doublet of doublets at *ca.* δ − 2.4 and δ − 4.5 ppm due to coupling to the two inequivalent phosphine ligands (the H–H coupling is not resolved). The broad resonance of the molecular hydrogen ligand in the non-classical tautomer overlaps the resonance of one of the two inequivalent hydrides of the classical complex; the resonance is broad due to fast dipole–dipole relaxation. We can demonstrate that the breadth of the resonance is due to a short T₂ by replacing H₂ with HD (Figure 4.17b, c). The gyromagnetic ratio of ²H is much smaller than that of ¹H so dipole–dipole relaxation is much less effective and the resonance narrows significantly. Furthermore, the ¹H resonance of the ligands is now a 1:1:1 triplet; J(HD) = 33.5 Hz as expected for coupling to ²H, which has I = 1 (Figure 4.17b). Typically, ¹J(HD) is between 25 and 34 Hz in molecular hydrogen complexes, whereas ²J(HD) in classical hydrides is less than 5 Hz.

Factors other than fast spin–spin relaxation—for example, chemical exchange and fluxional processes (see Chapter 6)—can also cause broadening of NMR lines, so a broad line alone is not enough to distinguish unequivocally molecular hydrogen from classical hydride complexes. Fortunately, no such ambiguity arises if we use spin–lattice relaxation to distinguish the two tautomers, since the protons of a molecular hydrogen ligand will clearly relax much faster than those of a classical hydride. Care is still needed in the interpretation of these results, since inspection of Figure 4.10 reveals that T₁ depends on both field and temperature, through the effect of temperature on the molecular correlation time. To eliminate these effects, it is essential that T₁ is measured at the same spectrometer magnetic field and that the value at the T₁ minimum is determined.

When this is done spin–lattice relaxation times of less than 10 ms (at 5.875 T, i.e. $^1H = 250$ MHz) indicate a molecular hydrogen ligand, whilst T_1s greater than 100 ms indicate classical hydrides.

Spectral editing using relaxation times

Differences in relaxation times can be used to resolve overlapping resonances if the species concerned relax at different rates. Inspection of Figures 4.13 and 4.15 reveals that after application of a 180° pulse, the magnetization relaxes back along the *z*-axis, *passing through zero*. This will occur after different times for species having different T_1s. This can be seen in the T_1 edited proton NMR spectrum shown in Figure 4.17d. If we invert the magnetization of all the protons in a mixture of [W(CO)$_3$(P^{i}Pr$_3$)$_2$(H$_2$)] and [W(CO)$_3$(P^{i}Pr$_3$)$_2$(H)$_2$], the resonances of the molecular hydrogen ligand, which are relaxing ten times faster than those of the classical hydrides, will pass through zero, whilst those of the classical hydrides have barely relaxed, so are still large (and negative). If the ^1H NMR spectrum of the excited system is measured at this point, there will be no resonance due to the molecular hydrogen ligand—its magnetization is zero—and a strong set of resonances, of negative intensity, for the classical hydride ligands (Figure 4.17d).

Contrast agents for magnetic resonance imaging

Magnetic resonance imaging is used in medicine, for example to detect tumours by visualizing soft tissue, or to study brain activity by localizing and monitoring blood flow. This is achieved by monitoring the proton spins in the water in cells or fat or blood following an excitation pulse.

Spatial localization of the signal is achieved by applying a magnetic field gradient which, in its simplest implementation, alters the magnetic field experienced by the protons and hence their resonance frequency *as a function of position* (see Chapter 1, equations 1.2 and 1.3). MRI uses this localization of the differences in magnetic properties (resonance frequency, relaxation rate, magnetic susceptibility) of the protons in tissues to construct an image of the sample (Figure 4.18).

Since the image is created from the NMR signal of water in the body, the quality of the image—the contrast between the tissue of interest and its surroundings—depends on the ability to distinguish the water signal in the tissue of interest from the background water signal. This can be achieved by the addition of paramagnetic transition-metal or lanthanide complexes to enhance the relaxation rate in one, or other, region of the sample.

In MRI rapid pulsing is used when acquiring the image; proton spins in slowly relaxing regions become saturated, while water in tissues with enhanced spin–lattice relaxation (short T_1) have time to relax before the next excitation pulse is delivered, i.e. the signal from these regions of the sample will have more intensity than that from regions without T_1 enhancement. Conversely, areas with short spin–spin T_2 values give lower signal intensity, since this diminishes the net

Dipole–dipole relaxation falls away as r^{-6}, and increases as the product of the squares of the gyromagnetic ratios of the relaxing pair of nuclei. The short distance between the two hydrogen atoms in a molecular hydrogen ligand results in very efficient dipole–dipole relaxation.

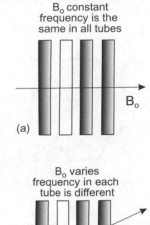
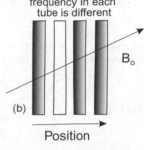

Figure 4.18 If the magnetic field is the same at all positions in the sample, all water protons will resonant at the same frequency. However, by applying a magnetic field gradient across the sample, the proton resonance frequency varies as a function of position Thus, in (a), if one tube is empty we cannot tell which one it is, whereas in (b) the protons in each tube will resonate at a different frequency and we can identify the empty tube by the absence of a signal at its resonance frequency. This is the idea behind magnetic resonance imaging; magnetic field gradients are used to encode the NMR signal as a function of position within the sample. Plotting the NMR signal as a function of this spatial encoding generates an image of the sample.

Relaxation is an important contribution to NMR line widths since this determines the lifetime of the excited state. Rearranging the Heisenberg equation gives $\nu_{1/2} = 1/\pi T_2$, $\nu_{1/2}$ = peak width at half height.

	R
Gadoterate	CH_2CO_2H
Gadoteridol	$CH(Me)CH_2OH$

Gadobutrol

Gadopentetate family

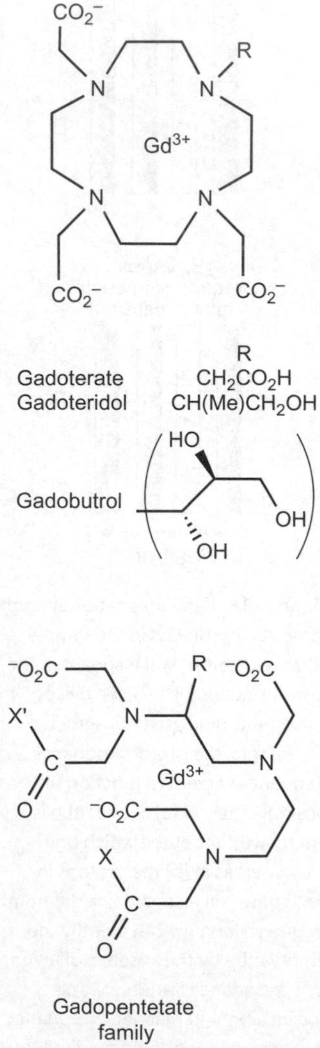

Figure 4.19 Gadolinium contrast agents used in MRI are normally complexes of polydentate and/or macrocyclic ligands to minimize release of the toxic metal ion into the body

magnetization in the x–y plane available for detection irrespective of the pulsing rate; i.e. unwanted signals are removed.

Most paramagnetic contrast agents approved for medical use are gadolinium complexes of tetraazacyclododecane-1,4,7,10-tetraacetate (DOTA) or closely related ligands (Figure 4.19). Lanthanide ions, especially gadolinium, have received the most attention because of the high magnetic moments that arise from unpaired 4f electrons, for example Gd^{3+} has an $^8S_{7/2}$ ground state and an effective magnetic moment around 8 μ_B. The symmetric S ground state of Gd^{3+} is important since it results in an electronic relaxation rate more in tune with relaxation of the proton nuclear spins, whereas water protons hardly feel the effects of ions such as Dy^{3+} or Ho^{3+}, which have higher magnetic moments but faster relaxing, asymmetric electronic ground states.

The ionic radius of Gd^{3+} (108 p.m.) is very close to that of Ca^{2+} (114 p.m.). Free gadolinium ions are, therefore, competitive inhibitors of physiological processes that depend on Ca^{2+}. Concerns have also been raised over the potential neurotoxicity of gadolinium; care must be taken in designing the ligands in these contrast agents to avoid release of highly toxic 'free' Gd^{3+} ions into the patient. All Gd-based MR contrast agents use chelating or macrocyclic ligands (Figure 4.19), which are essentially inert to substitution, in contrast to the kinetic lability normally associated with lanthanide ions. The toxicity of gadolinium has led to complexes of other paramagnetic transition metal ions such as manganese, the lanthanides Eu (III) or Dy (III), and the use of ferromagnetic iron oxide nanoparticles, being explored.

Recently, transition-metal complexes with slowly exchanging –NH or –OH protons on the ligand have been shown to have potential to alter tissue contrast via chemical exchange saturation transfer (CEST) of presaturated spins with bulk water. Paramagnetic ions such as Mn^{2+} Dy^{3+}, Tb $^{3+}$, Tm $^{3+}$, and Yb^{3+} combined with ligands having exchangeable OH or NH protons offer a double advantage, since the resonance to be presaturated will experience a large paramagnetic shift in its frequency, moving it clear of the bulk water resonance, thus improving the selectivity of presaturation. Such slow water exchange paramagnetic complexes are referred to as PARACEST agents. These compounds also have no need for a vacant co-ordination site at the metal, reducing the risk of deco-ordination of the metal from the ligand.

4.6 Summary

- Fourier transform NMR uses pulses of radiation to excite the nuclear spins in a sample. This allows for rapid acquisition of NMR spectra in the time domain enabling the spectrum to be measured many times; adding the results improves the signal-to-noise ratio dramatically (sample averaging).

- The vector model provides a simple means to visualize the effect of an excitation pulse on the NMR magnetization and the evolution of the magnetization vector in the rotating frame and can be used to understand simple pulse sequences.

- It is possible to design sequences of pulses to access specific information about the sample.

- Relaxation is an important consideration when using excitation pulses and sample averaging, since we must allow the system to return to equilibrium before starting the excitation sequence again.

- This is particularly important when using multiple-pulse sequences such as those discussed in Chapter 5.

- Relaxation can also provide valuable structural information and dynamic information about a sample that is not accessible through analysis of the chemical shifts and couplings.

- The application of pulse sequences and relaxation in advanced NMR techniques including 2-D spectroscopy (**COSY**, **HSQC**, **NOESY**, **EXSY**, etc.) is explored further in Chapter 5.

4.7 Exercises

1. Draw vector diagrams to illustrate the effect of a $90°$ $-x$ pulse on the ^{19}F magnetization of bromotrifluoromethane and its subsequent precession.

2. How will the picture in Q.1 change for trifluoromethane, i.e. when Br is replaced by H? For simplicity, assume the ^{19}F chemical shift is exactly on resonance, i.e. ignore the precession due to the chemical shift.

3. A doublet resonance was first excited by a $90°$ $-y$ pulse then allowed to precess for a time τ before a $180°$ x pulse was applied. The system was then allowed to evolve for a further τ. Draw vector diagrams to illustrate the effect of the pulses and explain where the echo will form.

4. An inversion recovery sequence was used to determine the $^{31}P\{^{1}H\}$ spin–lattice relaxation times of a series of palladium–phosphine complexes [PdCl$_2$L$_2$]. The 'zero-crossing' times, t, of the phosphorus resonances are given in Table 4.1. Calculate the T_1 values for each complex and suggest a reason for the faster relaxation observed for the trialkyl phosphines.

Table 4.1 $^{31}P\{^{1}H\}$ zero-crossing times for several [PdCl$_2$L$_2$] complexes

L	PPr$_3$	PBu$_3$	P(C$_6$H$_{11}$)$_3$	P(Me$_2$Ph)$_3$	P(MePh$_2$)$_3$	P(p-tol)$_3$
t/s	4.4	3	2.5	9.6	7.5	7.2

Source: From data in W. Bosch and P. S. Pregosin (1979), *Helv. Chim. Acta.*, **82**, 838–43.

Polarization transfer and 2-D NMR spectroscopy

5.1 Introduction

So far we have looked at experiments in which a simple 1-D spectrum is recorded and how NMR spectroscopy can be used to determine the structure of simple molecules. As molecular complexity increases interpretation of the NMR spectrum can become more challenging; it may be difficult to separate overlapping resonances or to assign coupling partners. We may be interested in the conformation of a molecule, i.e. how close together particular groups in the molecule are that are not directly bonded and don't show J coupling. Not all compounds have rigid structures; for example, rotation can occur about the bonds in a molecule, rings can spin or change conformation, and ligands can migrate between different sites. Or the nuclide we are interested in may be very insensitive and we need to increase its NMR signal. We can divide the NMR active metals on the basis of how easy it is to obtain useful structural information either via scalar couplings/satellites in the ligand spectra or from direct observation of the metal spectra. Thus ^{113}Cd, ^{119}Sn, ^{195}Pt, and ^{199}Hg are easily observed, ^{107}Ag, ^{109}Ag, and ^{103}Rh are particularly useful through their couplings (high natural abundance) but their low γ results in low sensitivity making direct detection difficult, while direct detection of ^{57}Fe, ^{99}Ru, ^{183}W, and ^{187}Os is extremely challenging due to a combination of the low γ and low natural abundance. Fortunately, it is possible to use polarization transfer from a high γ spin, such as protons or ^{31}P effectively to increase sensitivity in the NMR observation of a low γ metal.

5.2 Polarization transfer

Polarization transfer is the transfer of energy between spin populations and can be used to elucidate a particular property, for example the couplings between the nuclear spin of (an NMR-active) metal and its ligands or, as noted in the introduction, to make observation of an insensitive spin easier. Polarization transfer can also be used to monitor **dynamic processes**. Modern, multiple-pulse and polarization transfer NMR methods are the subject of this chapter, while dynamic processes are discussed in Chapter 6.

INEPT

INEPT (Insensitive Nucleus Enhancement by Polarization Transfer) was one of the first polarization transfer experiments developed by Morris and Freeman in the late 1970s. In those experiments, a sequence of r.f. pulses was used to transfer polarization from the sensitive protons to carbon, thereby increasing the sensitivity in ^{13}C NMR. Since that time instrumentation has become available enabling such multiple-pulse techniques to be used to increase the NMR signal intensity of insensitive metal nuclei, for example ^{57}Fe, ^{103}Rh, ^{109}Ag, and ^{187}Os, that are coupled to abundant sensitive spins such as ^{1}H and ^{31}P. Even ^{13}C can be used as the sensitive nucleus if the compound is enriched in ^{13}C. Amongst 1-D multiple-pulse experiments, INEPT (Figure 5.1) and its derivatives are perhaps the most important. The effect of the sequence is to increase the population difference between the lower and upper levels of the S spins (Figure 5.2), thus increasing the NMR signal intensity of the S spins.

We saw in Chapter 1, section 1.4, that the intensity of the NMR signal depends on this population difference, hence the INEPT experiment increases the intensity of the NMR signal of the insensitive nucleus. The increase in intensity is equal to the ratio of the gyromagnetic ratios of the two nuclei (Table 1.2). In this way, the intensity of signals from insensitive nuclei, i.e. ^{59}Fe, ^{103}Rh, or ^{109}Ag, can be increased by more than an order of magnitude by transfer of polarization from ^{1}H using INEPT affording up to two orders of magnitude improvement in signal to noise.

INEPT thus provides an excellent tool for recording the NMR spectra of low γ metals, the large enhancement in signal intensity resulting in significant saving in experimental time; Figure 5.3 compares the ^{109}Ag{^{1}H} spectra of

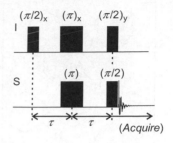

Figure 5.1 The INEPT pulse sequence transfers energy between the spins in a coupled system. τ is set to $1/(4J)$. (Pulses are represented by shaded blocks.)

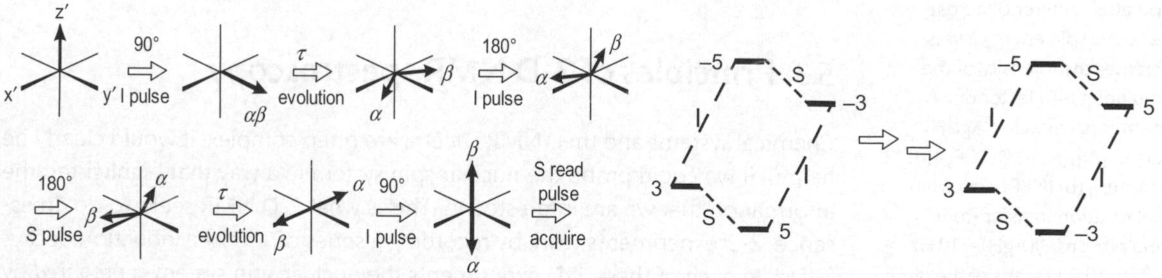

Figure 5.2 The effect of the INEPT pulse sequence on the I spin populations of a coupled pair of spins, I–S, can be described using the vector model described in Chapter 4; only the I spin vectors are shown. The first 90° pulse translates the I magnetization into the xy-plane. After a delay $\tau = 1/(4J)$ the two halves of the I doublet have precessed to be 90° apart. The 180° I pulse flips the spins in the xy-plane about the y-axis whilst the simultaneous 180° S pulse interchanges the α and β spin labels. After a further delay of $1/(4J)$ the two vectors representing the two lines of the doublet are 180° out of phase and lie along the x-axis. A second 90° pulse now aligns the I magnetization along +/−z. The populations of the various energy levels are shown in the diagram on the right. The effect of the pulse sequence is to invert the population difference between the upper and lower energy levels of one line of the I spin system. The effect of this inversion is to increase the magnitude of the population differences in the S spin states and results in an increase in the intensity of the insensitive nucleus transitions. Populations shown here are for an HC spin system; I and S spectroscopic transitions are shown by long and short dashed lines, respectively.

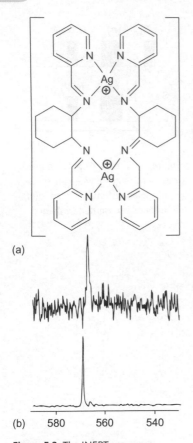

(a)

(b)   580   560   540

Figure 5.3 The INEPT sequence
inverts the populations of one line
of the *I* spin multiplet, increasing
the population difference across
the insensitive spin energy levels.
This increases the intensity of the
S spin multiplet by a factor of γ_I/γ_S.
(a) Conventional direct ^{109}Ag{^{1}H}
spectrum of $[Ag_2(N_4)_2](O_3SCF_3)_2$,
(55,000 scans); (b) INEPT spectrum
using polarization transfer from
the imine protons, ^3J(AgH) ~ 10 Hz,
γ_H/γ_{Ag}, ~ 20, 10,000 scans of the same
solution.

Source: Adapted with permission from G.
Van Koten et al. (1983), *Inorg. Chim. Acta*, **79**,
206–07. Copyright © 1983 Elsevier B.V.

$[Ag_2(N_4)_2](O_3SCF_3)_2$ acquired using the INEPT pulse sequence with a conventional pulse-acquire spectrum. One-fifth the experimental time was required to achieve more than an order of magnitude improvement in signal to noise, equivalent to a *ca.* 500-fold saving in experimental time.

The distortions introduced by the basic INEPT sequence can be removed using the refocused INEPT experiment, to obtain an NMR spectrum that looks exactly like a conventional NMR spectrum. However, care must be taken in using intensity information from the INEPT family of experiments, since this no longer depends simply on the number of nuclei present, but also on the efficiency and degree of enhancement.

Unlike inverse detection methods such as heteronuclear multiple quantum correlation (HMQC; see 5.3 'Principles of 2-D NMR spectroscopy' below), coupling to the sensitive spins can be retained; this can be particularly useful in determining the number of donor spins, and hence ligands attached to the metal centre. For example, although it is possible to extract coupling information from the $^{117/119}$Sn and ^{195}Pt satellites in the ^{1}H spectrum of *cis*-[Pt(PPh$_3$)$_2$(C≡CSnMe$_3$)(SnMe$_3$)], this can be a formidable task due to the complexity of the ^{1}H spectrum. On the other hand, interpretation of the ^{119}Sn,^{1}H{^{1}H} INEPT spectrum (Figure 5.4) is relatively straightforward. Two different ^{119}Sn environments are clearly seen, as is the doublet-of-doublets structure due to coupling of the downfield ^{119}Sn resonance (δ ~ –61 ppm) to the two different PPh$_3$ ligands (the small and larger splitting ^2J(^{119}Sn^{31}P) = 110 and 1606 Hz, corresponding to *cis* and *trans* coupling, respectively). The tin centres show ^{195}Pt satellites with very different coupling constants consistent with one tin being directly bonded to platinum and the other not, ^1J(^{119}Sn^{195}Pt) = 8140, and ^3J(^{119}Sn^{195}Pt) = 279 Hz. The observation of $^{117/119}$Sn satellites on the upfield ^{119}Sn resonance (δ ~ –88.9 ppm, ^4J(^{119}Sn^{117}Sn) = 17.5 Hz) (Figure 5.4 insert) confirms the presence of two tin centres in the molecule.

5.3 **Principles of 2-D NMR spectroscopy**

Chemical systems and their NMR spectra are often complex; it would clearly be helpful if we could probe the nuclear spin system in a way that highlighted the information that we are interested in. This is what 2-D NMR seeks to do. In essence, 2-D experiments work by recording a series of 1-D free induction decays (FIDs). In each of these 1-D experiments the nuclear spin system is *prepared* by a series of excitation pulses and delays. One of these delays, during which the magnetization evolves under the influence of the NMR property of interest (e.g., the time allowed for exchange to occur in an EXSY (exchange experiment)), is sequentially increased (incremented) between one 1-D experiment and the next (Figure 5.5). In this way we create two different time dimensions; one is 'real', the acquisition time of the 1-D FID (called F$_2$), the other, (called F$_1$), is 'synthesized' by the incrementing delay in the pulse sequence (Figure 5.5). We can now do Fourier transformations (FTs) in both time dimensions to produce a 2-D map in which cross-peaks occur that relate the chemical shifts in the 1-D spectrum (F$_2$) to the property selected by the pulse sequence in the second (F$_1$).

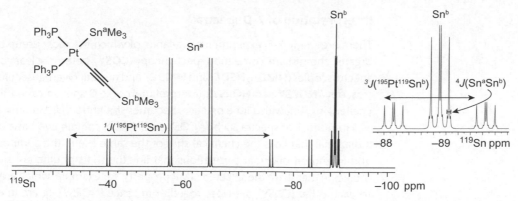

Figure 5.4 ^{119}Sn NMR spectrum of *cis*-[Pt(PPh$_3$)$_2$(C≡CSnMe$_3$)(SnMe$_3$)], measured using the refocused INEPT pulse sequence with ^{1}H decoupling. The expected splitting of the central resonances is observed together with the ^{195}Pt satellites. The expanded region of the ^{119}Sn (≡CSn) resonance shows $^{117/119}$Sn satellites due to ^4J(Sn^a,Sn^b).

Source: Simulated from data in M. Herberhold et al. (1998), *Chem. Eur. J.*, 4, 1027–32.

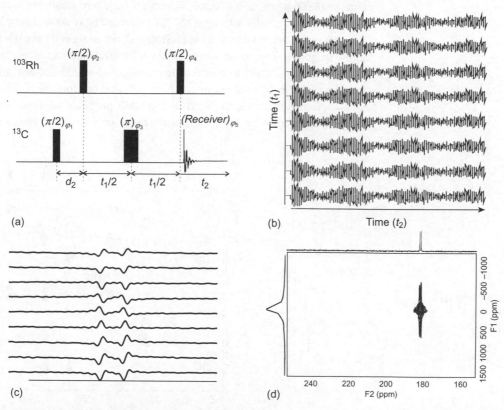

Figure 5.5 In a 2-D experiment the nuclear spins are first prepared by a series of excitation pulses and mixing delays (a). One of the mixing delays is incremented to generate a second time dimension. Shown here is the basic HMQC experiment using ^{13}C as the *I* spin and ^{103}Rh as the *S* spin; t_1 is the incrementing delay. The incrementing delay, t_1, creates a second time dimension, F1, (b), that can be used to relate two separate NMR properties to each other, for example chemical shifts of two different spins or chemical shift and scalar coupling. Fourier transformation in both dimensions (c) gives a map (d), in which cross-peaks occur that relate these two NMR properties.

Interpretation of 2-D spectra

There are many 2-D experiments available, of which the most useful to the inorganic chemist are: correlation spectroscopy (**COSY**), heteronuclear correlation spectroscopies (**HMQC**, **HSQC**, and **HMBC**), and nuclear Overhauser effect spectroscopy (**NOESY** and **HOESY**). Interpretation of 2-D spectra can at first seem challenging, but with a little practice, becomes straightforward in most cases.

Homonuclear spectra such as COSY and NOESY contain two basic elements: a diagonal that links the chemical shift of the same site in the compound in F_2 and F_1 and off-diagonal correlations that link the shifts of different sites in the molecule that share the selected NMR property; for example, which are coupled to each other (COSY), are close together in space (NOESY), or are in exchange with each other (EXSY). Correlations will thus consist of four elements: two on the diagonal and two off the diagonal. Identifying cross-peaks and diagonals that 'belong together' is simply a case of drawing a box (Figure 5.6b). From a cross-peak move across to the diagonal, then down to the other cross-peak, back to the other diagonal element, and finally back up to the original cross-peak.

Artefacts are frequently observed in NOESY spectra due to chemical exchange (EXSY peaks, see Chapter 6, section 6.2), zero quantum artefacts, and COSY peaks. For small molecules, NOESY peaks will be weak and have the opposite sign to the diagonals, whereas EXSY and COSY peaks will have the same sign as the diagonals. Zero-quantum artefacts will show dispersive character.

Heteronuclear spectra do not contain a diagonal, so correlations are found by looking across from each chemical shift on F_1 and up from each chemical shift on F_2. If the two shifts are related by the NMR property selected by the pulse programme, a cross-peak will exist where these lines intersect (Figure 5.6a).

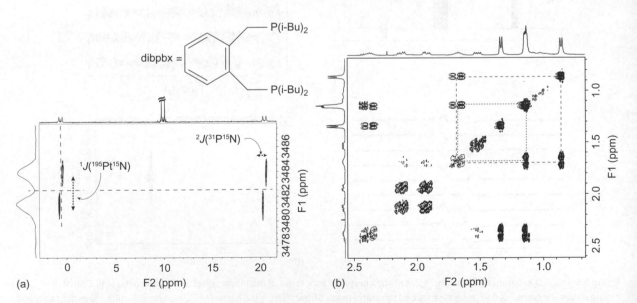

Figure 5.6 (a) $^{31}P,^{195}Pt\{^1H\}$HMQC spectrum of $[Pt(PPh_3)(^{15}N\text{-pyridine})Cl_2]$ showing assignment of the $^{31}P,^{195}Pt$ correlations; couplings to ^{15}N are indicated. (b) *iso*-propyl region of the phase sensitive, double quantum filtered ^1H COSY of $[Pd(dibpbx)Cl_2]$ showing the assignment of the correlations due to the coupling between the methyne and (inequivalent) methyl groups in one P(*iso*-butyl)$_2$ unit.

Where couplings to third nuclei exist, these may appear in either F_1 or F_2, depending on the particular experiment, or may have been removed by the pulse sequence or the use of decoupling.

5.4 Applications of 2-D NMR to inorganic systems

COSY

A COSY spectrum can be looked on, at least as far as the information it contains, as a 2-D version of a 1-D selective decoupling experiment (Figure 5.6). In the 2-D map, diagonal signals correspond to the individual chemical shifts in the 1-D NMR experiment, while off-diagonal, aka cross-peaks link coupled nuclei.

Consider, for example, the macrocyclic diphosphine, **1** (Figure 5.7), which provides an example of through-space coupling *via* the overlap of the phosphorus lone pairs. Two distinct chemical shifts are seen in the $^{31}P\{^1H\}$ spectrum as expected; unexpectedly, however, these appear as well-resolved doublets with $J = 67$ Hz. At first sight, the phosphorus centres appear too far apart to couple to each other, being separated by eight covalent bonds; however, there are no other possible coupling partners in the macrocycle. A 2-D $^{31}P,^{31}P$-COSY clearly demonstrates that the coupling is indeed phosphorus–phosphorus—cross-peaks

We are familiar with scalar couplings between nuclei linked by a covalently bonded network. Less commonly, coupling is observed between nuclei that are not linked by a covalent bonded network or the network involves too many bonds for scalar coupling via the network to be expected. When non-bonded nuclei are brought close together by the molecular conformation J coupling may still occur through the Fermi contact term, even though no covalent bond exists; these are called through-space couplings. Of relevance to inorganic chemistry is coupling through overlapping lone pairs which can be seen, for example in the case of unsymmetrical diphosphine systems; see B. A. Chalmers et al. (2018), *Inorg. Chem.*, **57**, 3387–98 and J. C. Hierso (2014), *Chem. Rev.*, **114**, 4838–67 for more information.

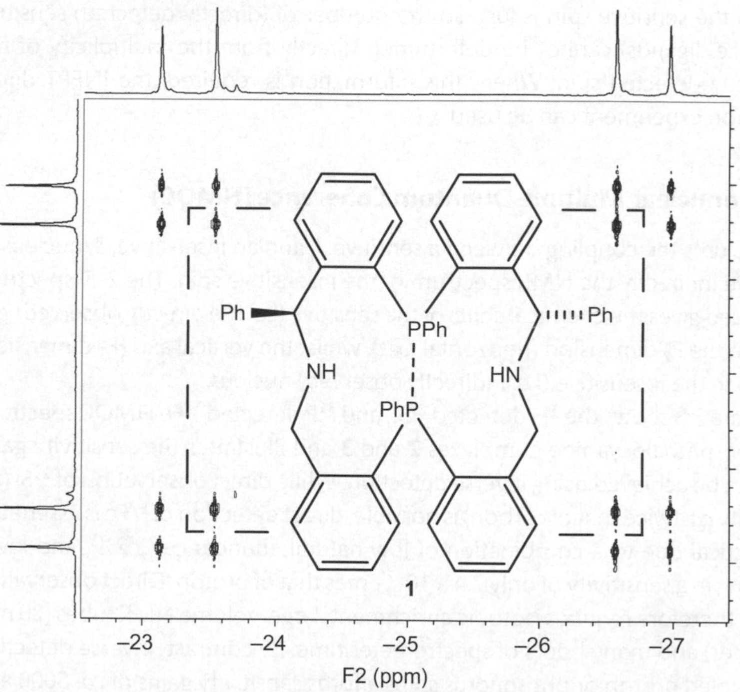

Figure 5.7 Through-space scalar coupling between the phosphorus nuclei, indicated by dotted lines, is seen in the $^{31}P,^{31}P$-COSY of macrocycle **1** and results from overlap of the phosphorus lone pairs

Source: N. Ritchie, J. A. Iggo, and J. L. Xiao, unpublished.

are seen that link the two phosphorus chemical shifts. The through-space coupling originates from the overlap of the phosphorus lone pairs, which provides a pathway for transmitting spin information between the two phosphorus nuclei.

Inverse (indirect) detection

In the INEPT and COSY experiments above, we record the nucleus we are studying directly. However, a significant gain in sensitivity can be achieved if we observe the spectrum of the more sensitive nucleus, transferring polarization from the coupled, more sensitive spin to the lower γ spin of interest and back again. The NMR spectrum of the more sensitive spin is thus modulated by the coupling to the insensitive spin. Such experiments (e.g. HSQC, HMBC, and HMQC) are called inverse detected and are familiar to the organic chemist, where they are used primarily to reveal proton–carbon connectivities; in inorganic chemistry it is often the boost to sensitivity that is of prime importance.

The sensitivity gains afforded by HMQC and HSQC have revolutionized the NMR spectroscopy of low γ heteronuclei; for example it is possible to record ^{31}P-detected ^{103}Rh HMQC spectra in minutes, in contrast to direct detection of the same sample which might take many hours. Inverse detection, however, cannot compensate for the low natural abundance of a spin of interest since only the satellites from coupling to the low abundance spin are observed. The principle disadvantage of these inverse detection experiments is that coupling information to the sensitive spin is lost, so the number of (directly detected) sensitive spins (i.e. ligands) cannot be determined directly from the multiplicity of the indirectly detected spin. Where this information is required, the INEPT direct detection experiment can be used.

Heteronuclear Multiple Quantum Coherence (HMQC)

HMQC uses the coupling between a sensitive, I, and an insensitive, S, nucleus to observe indirectly the NMR spectrum of the insensitive spin. The 2-D spectrum produced gives the chemical shifts of the sensitive (i.e. the directly observed) nucleus in the F_2 dimension (horizontal axis), whilst the vertical axis (F_1 dimension) relates to the insensitive (i.e. indirectly observed) nucleus.

Figure 5.8 shows the ^{1}H detected ^{29}Si, and ^{31}P detected ^{57}Fe HMQC spectra of the iron-phthalocyanine complexes **2** and **3** and illustrates the sensitivity gains that can be achieved using inverse detection. While direct observation of ^{29}Si (receptivity *ca.* twice that of carbon) is possible, direct detection of ^{57}Fe is essentially impractical due to a combination of low natural abundance (2.2%), and low γ resulting in a sensitivity of only 7.4×10^{-7} times that of proton. Direct observation of ^{57}Fe therefore requires isotopic enrichment, large-volume NMR tubes (20 mm diameter) and many hours of spectrometer time. By contrast, inverse detection via coupled proton or phosphorus spins affords sensitivity gains of *ca.* 5000 and 550 times, respectively, making observation of ^{57}Fe a practical proposition; thus the HMQC spectra of Figure 5.8 were acquired in less than three hours.

The sensitivity of a heteronuclear NMR experiment is related to the gyromagnetic ratios of the starting and the detected spins and a relaxation term (equation 5.1).

$$S/N \propto \gamma_{start}\gamma_{detect}^{3/2}$$

where γ_{start} and γ_{detect} are the gyromagnetic ratios of the initially excited and detector spins, respectively. The sensitivity advantage of HMQC and HSQC is obvious from equation 5.1, $\gamma_{start} = \gamma_I$, $\gamma_{detect} = \gamma_I$, hence the sensitivity enhancement is $\sim (\gamma_I/\gamma_S)^{5/2}$ (cf. a direct, polarization transfer experiment such as INEPT, $\gamma_{start} = \gamma_I$, $\gamma_{detect} = \gamma_S$, sensitivity enhancement $\sim \gamma_I/\gamma_S$). Furthermore, T_1 of the I spin is commonly shorter than that of the S spin, so the whole pulse sequence-acquire experiment can be repeated more frequently, allowing more scans to be acquired in the same time, further improving the signal to noise.

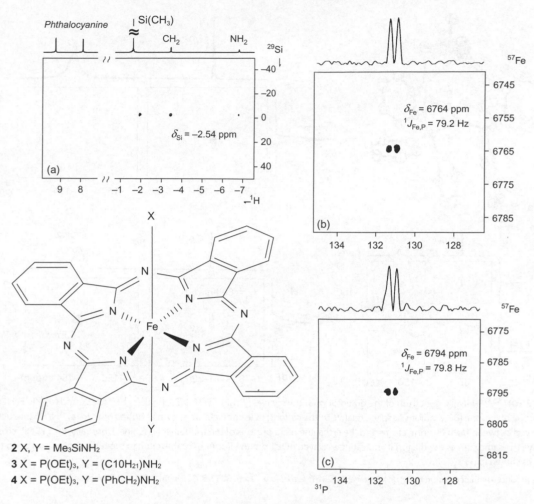

Figure 5.8 (a) ^{1}H,^{29}Si; and (b, c) ^{31}P,^{57}Fe-HMQC spectra of iron phthalocyanine complexes illustrating the sensitivity gains from indirect detection of low γ nuclei afforded by inverse detection. The F_1 projections (not shown) each show a 'singlet' at the S spin chemical shift.

Source: Adapted from P. Ona-Burgos et al. (2010), *Dalton Trans.*, 6231–8. Copyright 2010 The Royal Society of Chemistry.

Of course, HMQC experiments can also be used to establish connectivity in inorganic compounds (Figure 5.9). Thus, a ^{19}F,^{31}P–HMQC experiment was used to assign the resonances in the ^{19}F and ^{31}P NMR spectra of the difluoroborenium species [(Ph$_3$P)$_2$C–BF$_2$]$^+$[BF$_4$]$^-$. A correlation is seen between the upfield (δ_F = –81.1 ppm) ^{19}F resonance and that of the two magnetically equivalent phosphorus atoms (δ_P *ca.* 24.7 ppm), whereas no correlation is seen between phosphorus and the ^{19}F resonance at –153.4 ppm (Figure 5.9b). The absence of a ^{19}F–^{31}P correlation, the breadth of the ^{19}F (broadened by the quadrupole of boron) resonance, and its chemical shift all confirm that this latter resonance is due to the tetrafluoroborate anion.

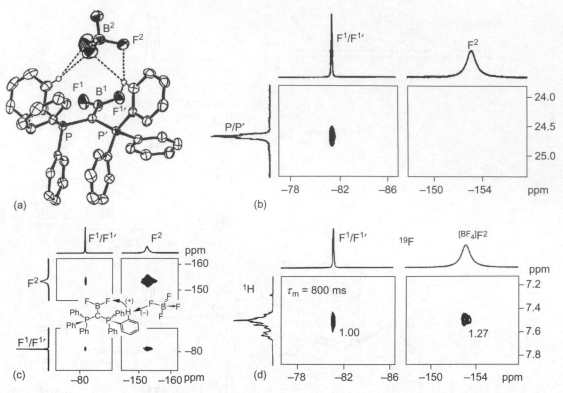

Figure 5.9 (b) ^{19}F,^{31}P-HMQC spectrum of the difluoroborenium species [(Ph$_3$P)$_2$C–BF$_2$]$^+$[BF$_4$]$^-$ (a); (c) ^{19}F NOESY; and (d) ^{19}F,^{1}H HOESY spectroscopy places the [BF$_4$]$^-$ anions in space relative to the cations; nOe contacts are shown as dotted lines in (a). ^{1}H/^{19}F interactions are observed between both fluorine atoms and the *ortho* and *meta* protons of the triphenylphosphine. Unusually, the HOESY cross-peaks have the same sign as the diagonal, indicating a relayed nOe from the fluorines of the anion to the *ortho* proton and on to the BF$_2$ fluorine, as shown in (c).

Source: Reproduced with permission from J. E. Munzer et al. (2016), *Eur. J. Inorg. Chem.*, 3852–8. Copyright © 2016 John Wiley and Sons.

Heteronuclear Single Quantum Coherence (HSQC)

HMQC has the advantage that the pulse sequence is short, so loss of signal due to relaxation during the pulse sequence is minimized; however, it is not without problems: the basic sequence gives skew line shapes, reducing the resolution; multiple quantum transitions result in passive couplings *between sensitive* spins appearing in the *insensitive spin spectrum* (Figure 5.10). If the insensitive spin has high abundance, for example ^{103}Rh or ^{109}Ag, multiple spin flips of the insensitive spin can occur, resulting in unexpected correlations, or no signal at all at the expected shift. HSQC uses a much longer pulse sequence (a dozen or more pulses as opposed to four in the basic HMQC sequence) so is more sensitive to mis-setting of the pulses and delays and to relaxation during the pulse sequence. However, HSQC is based on single quantum transitions, thus avoiding many of the problems of HMQC above and has the advantage that it gives much narrower resonances/cross-peaks (Figure 5.10), so is favoured where the spectrum of the sensitive spin is crowded. HSQC does, however, have its own 'wrinkles'; for

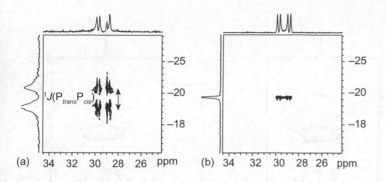

Figure 5.10 Phase-sensitive $^{31}P,^{103}Rh\{^1H\}$ HMQC (a) and HSQC (b) spectra of Wilkinson's catalyst, [Rh(PPh$_3$)$_3$Cl], illustrating the splitting of the resonances in F$_1$ (^{103}Rh) by passive ^2J(PP) couplings between the *cis* and *trans* phosphine ligands. Only the *trans*-PPh$_3$ region is shown. Modern variants of the HMQC sequence may include a refocusing pulse to remove this passive coupling.

example, in the $^{31}P\{^1H\},^{109}$Ag HSQC spectrum of **5** (Figure 5.11), couplings to silver are seen in both F$_1$ and F$_2$. In F$_2$ the true $^1J(^{31}P^{107}$Ag) and $^1J(^{31}P^{109}$Ag) coupling constants are seen but in F$_1$, the correlations are split by twice $^1J(^{31}P^{107}$Ag) as a result of the workings of the pulse sequence. This turns out to be a useful artefact of the pulse sequence in this case, since it confirms that two phosphorus spins couple to the silver, confirming that the dimer remains intact in solution.

Heteronuclear Multiple Bond Correlation (HMBC)

The HMBC pulse sequence is routinely used in organic chemistry to observe long-range proton–carbon couplings while excluding one-bond correlations. HMBC suffers from low sensitivity because of relaxation during the long delay inserted to filter out, large, short-range couplings; however, with the exception of C–H couplings in the backbone of ligands, this is rarely an issue faced in inorganic systems. Nevertheless, HMBC is occasionally used instead of HSQC or HMQC to detect 2-D spectra where the coupling constant between the I and S spins is small, although it offers no obvious advantage.

NOESY and HOESY

Nuclear Overhauser effect (nOe) measurements (Chapter 4, section 4.4) use dipolar relaxation between nuclear spins that are close together in space, i.e. separated by less than 5Å to establish molecular conformation. The power of nOe is that the spins only need to be close together, they do not need to be covalently bonded; nOes can thus be used to determine non-covalent ligand

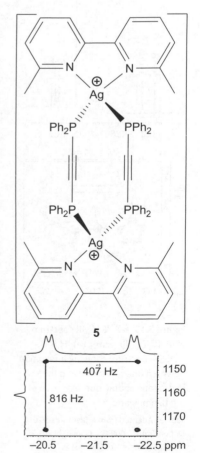

Figure 5.11 The $^{31}P\{^1H\},^{109}$Ag HSQC spectrum of **5** shows couplings to both silver isotopes in the ^{31}P dimension. In the ^{109}Ag dimension the splitting of the correlations is twice the coupling constant—an artefact of the pulse sequence. The F$_1$ projection, which does not show coupling of silver to phosphorus, was obtained by recording the $^{31}P\{^1H\},^{109}$Ag HSQC spectrum with ^{31}P refocusing pulse during the evolution period.

Source: Adapted from S. Keller et al. (2018), *Dalton Trans.*, **47**, 946. Copyright 2018 The Royal Society of Chemistry.

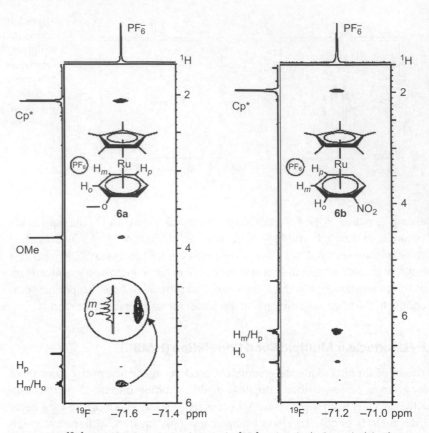

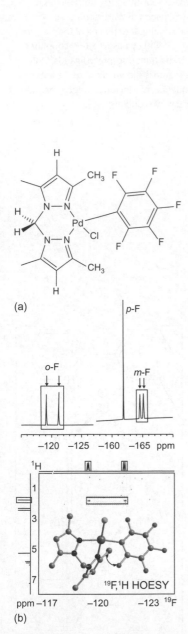

Figure 5.12 $^{19}F,^1H$ HOESY spectroscopy locates the $[PF_6]^-$ anions in the ion pairs lying between the *ortho* and *meta* protons in **6a** but between the *meta* and *para* protons in **6b**

Source: Adapted with permission from A. Moreno et al. (2008), *Chem. Eur. J.*, **14**, 5617–29. Copyright © WILEY-VCH Verlag GmbH & Co. KGaA, Weinheim.

Figure 5.13 (a) ^{19}F NMR spectrum of [Pd(C$_6$F$_5$)Cl(bpzm)]; (b) $^1H,^{19}F$– HOESY spectrum showing interaction between one *o*-fluorine and the hydrogens of only one Me group of the bzpm ligand

Source: Adapted from F. Blank et al. (2010), *Dalton Trans.*, 3609–19. Copyright 2010 The Royal Society of Chemistry.

binding sites, e.g. in proteins and ion pairs (Figures 5.9c,d, 5.12). nOe measurements can be performed either by a series of 1-D measurements or by recording a single, 2-D NOESY spectrum. nOes can exist between similar spins, for example $^1H,^1H$–NOESY, or between different spins, e.g. $^1H,^{19}F$–HOESY (heteronuclear Overhauser effect spectroscopy). One of the major disadvantages of NOESY is its intrinsically low sensitivity, requiring abundant spins as the 'donor', i.e. ^1H, ^{19}F for detection.

Consider the structure of the pentafluorophenyl palladium(II)-*bis*-(3,5-dimethylpyrazol-1-yl)methane (bzpm) complex shown in Figure 5.13. The ^1H NMR spectrum can be assigned and the hydrogen–carbon connectivities established by combining 1-D ^1H and ^{13}C NMR, and 2-D HMQC and COSY measurements. The bzpm methyl groups are inequivalent, as are the *ortho*-fluorines of the C$_5$F$_5$ ligand; two CH$_3$ resonances and two *ortho*-F resonances are observed in the ^1H and ^{19}F NMR spectra, respectively (Figure 5.13a) and two for *o*-F (^{19}F spectrum) centres. $^1H,^{19}F$-HOESY allows the 3-D structure of the complex to be established; through-space HOESY interactions are observed between only one of the *ortho*-fluorines of the C_5F_5 ligand and one of the methyl groups of bzpm ligand, indicating the bzpm ligand is not flat but folded at the CH_2 carbon, Figure 5.13b placing one pyrazolyl unit *trans* to the C_5F_5 ligand and the other *cis*.

5.5 Hyperpolarization techniques

So far we have looked at how the spin system can be manipulated by pulses, taking advantage of scalar coupling between a sensitive (high γ, I) and an insensitive (low γ, S) spin to increase the sensitivity of the NMR experiment. Another approach is to increase the 'initial' polarization of the nuclear spins before the NMR experiment. Three methods of doing this that are attracting interest in

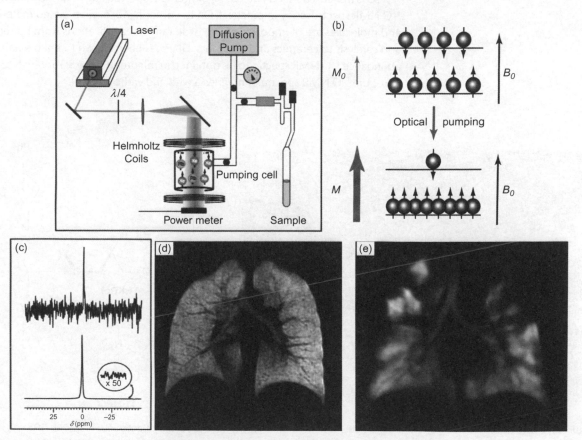

Figure 5.14 (a) Schematic of an optical pumping apparatus used to hyperpolarize ^{129}Xe; (b) laser excitation pumps the ^{129}Xe spins to the lower energy level, dramatically increasing the population difference across the NMR transition; (c) ^{129}Xe NMR spectrum of xenon gas, obtained with (bottom) and without (top) laser polarization. The lower spectrum reflects a signal enhancement of ~55,000; (d) and (e) ventilation images of healthy and cystic fibrosis-affected human lung acquired using hyperpolarized ^{129}Xe.

Sources: Adapted with permission from B. M. Goodson (2002), *J. Magn. Reson.*, **155**, 157–216. Copyright © 2002 Elsevier Science (USA); J. P. Mugler III and T. A. Altes (2013), *J. Magn. Reson. Imaging*, **37**, 313–31. © International Society for Magnetic Resonance in Medicine. See also E. Weiland et al. (2016), *Micropor. Mesopor. Mat.*, **225**, 41–65.

inorganic chemistry are: the use of *para*-hydrogen; the use of lasers to hyperpolarize, for example ^{129}Xe (dubbed spin-exchange optically pumped, or OPNMR), and the use of microwaves to transfer the much larger electron spin polarization to the nuclear spins—dynamic nuclear polarization or DNP. Hyperpolarization of xenon affords up to 16,000-fold enhancement of the ^{129}Xe NMR signal and has been used for example in medical NMR imaging and to study pore size in microporous materials, taking advantage of the high sensitivity of the xenon chemical shift and relaxation to its environment. Although the principles of DNP have been known for decades, it is only just beginning to re-emerge as a technique to study inorganic systems. Although the details of these techniques are beyond the scope of this book, the basic ideas of ^{129}Xe OPNMR and of DNP are illustrated in Figures 5.14 and 5.15; the interested reader is referred to the articles cited in the captions.

^{17}O is the least abundant (0.037%) oxygen isotope, making natural abundance ^{17}O NMR spectroscopy impractical. On the other hand, oxygen-based materials and molecules are of importance in a wide range of applications from batteries and catalysis to ceramics and polymers. DNP has been shown to enable natural abundance ^{17}O NMR spectra to be obtained in minutes (Figure 5.16, which compares the ^{17}O NMR spectra of magnesia with and without DNP).

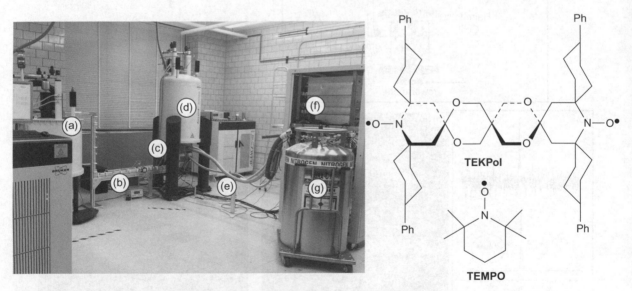

Figure 5.15 The Bruker 9.4-T/263-GHz DNP solid-state NMR spectrometer at the Ames Laboratory. Continuous-wave 263-GHz microwaves from the gyrotron (a) are transmitted via a waveguide (b) to the magic angle spinning probe (c) in the 9.4-T NMR magnet (d). The sample is held at a temperature of ~100 K by cold nitrogen gas (e, f, g). The microwaves, on or off, and microwave power are controlled by the PC (h). The sample is a frozen glass containing an organic radical in which the material of interest is suspended. TEMPO and TEKPOL are commonly used radicals. Polarization is transferred from the electron spin of the organic radical to the nuclear spins of the sample. DNP-enhanced ^{1}H polarization can then be transferred to lower γ heteronuclei such as ^{13}C, ^{15}N, or ^{17}O by using cross-polarization (CP) or other techniques to enhance the sensitivity of NMR experiments by a factor of up to 658.

Source: Reproduced with permission from L. Zhao et al. (2017), *Mag. Reson. Chem.*, **56**, 583–609. Copyright © 2018, John Wiley and Sons.

Figure 5.16 ^{17}O is the least abundant (0.037%) oxygen isotope, making natural abundance ^{17}O NMR spectroscopy impractical. On the other hand, oxygen-based materials and molecules are of importance in a wide range of applications from batteries and catalysis to ceramics and polymers. DNP allows natural abundance ^{17}O NMR spectra to be obtained in minutes. Compare the ^{17}O spectra of MgO at natural abundance obtained (a) without DNP; (b) with DNP; and (c) with DNP and double-frequency sweep, which illustrate the sensitivity gains that can be achieved by transferring polarization from electrons to the nuclear spins. The experiment time was 16 min.

Source: Adapted with permission from F. Blanc et al. (2013), *J. Amer. Chem. Soc.*, **135**, 2975–8. Copyright © 2013 American Chemical Society.

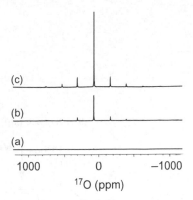

para-Hydrogen for Sensitivity Enhancement (PHIP)

In contrast to OPNMR and DNP, *para*-hydrogen experiments are easily implemented; a basic set-up requires little more than a lecture bottle of hydrogen gas and a dewar of liquid nitrogen (in addition to an NMR spectrometer) to implement.

The nuclear spins in H_2 can be aligned either parallel (both 'up') or antiparallel (one up, one down) giving rise to four spin states (Figure 5.17a). The populations of the different spin states are in equilibrium. If H_2 is cooled to very low temperatures in the presence of a suitable catalyst, *para*-hydrogen is formed, in which the $\alpha\beta$–$\beta\alpha$ spin state becomes dominant (Figure 5.17b); the nuclear spins are hyperpolarized. When *para*-hydrogen is reacted with a metal

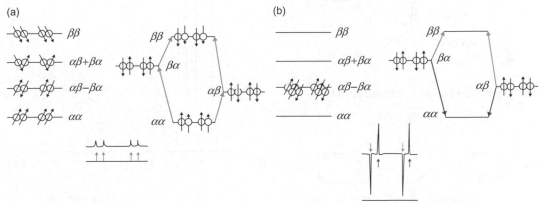

(a)
Spin states of "normal" H_2
All energy level populations are at equilibrium
Oxidative addition of "normal" H_2 to a metal complex gives a complex with equilibrium spin state populations
Spectrum has "normal" intensity

(b)
Spin states of hyperpolarized *para*-H_2
Populations of the $\alpha\beta$–$\beta\alpha$ energy level is increased at the expense of other spin states
Oxidative addition of *para*-H_2 to a metal complex gives a complex with non-equilibrium spin state populations
Spectral intensity is greatly increased

Figure 5.17 The intensity of an NMR spectrum depends on the population difference of the nuclear spin energy levels, Chapter 1.4. In a *para*-hydrogen experiment, the population of the $\alpha\beta$–$\beta\alpha$ spin state is increased. When this non-equilibrium population of spin states is carried over to the product of the hydrogen addition, a greatly enhanced signal intensity in the ^1H NMR spectrum of the hydride product results.

Source: Figures used courtesy of Prof. S. Duckett, University of York.

complex, provided the hydride ligands in the product are inequivalent, a hyper-polarized metal-dihydride complex is formed and a dramatic increase—up to a thousand-fold or greater—is seen in the intensity of the NMR signals which now appear in anti-phase, i.e. one up, one down. This process is called PHIP—*para*-hydrogen-induced polarization (Figure 5.17b).

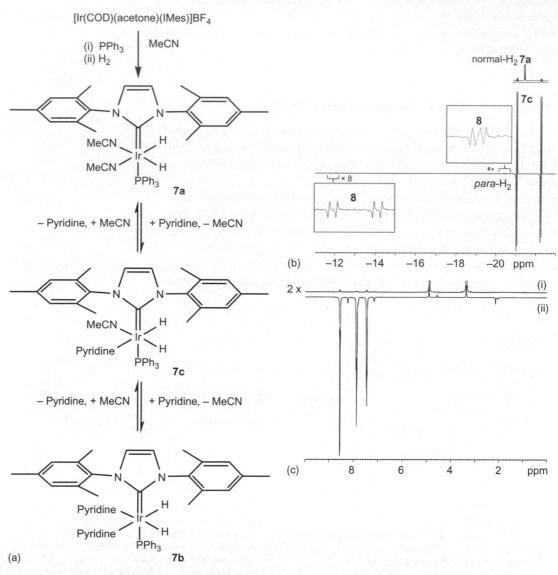

Figure 5.18 (a) Preparation of **7a–7c** from **2**. (b) Hydride region the ^{1}H NMR spectra of **7a** and **7c**, illustrating the PHIP intensity enhancement of the resonances of the inequivalent hydride ligands in **7c**. Note the resonances of traces of *cis,cis*-[Ir(H)₂(NCMe)(IMes) (PPh₃)₂]BF₄ **8**, which is completely undetectable using normal H₂, can just be seen in the PHIP spectrum. (c) ^{1}H NMR spectrum of a solution of **9** (the tricyclohexylphosphine analogue of **7c**) and pyridine on bubbling (i) normal; and (ii) *para*-hydrogen through the solution; note the dramatic increase I intensity of the pyridine resonances between 7 and 9 ppm in the SABRE spectrum.

Source: Adapted from M. Fekete et al. (2013), *Inorg. Chem.*, **52** 13453–61. Copyright © 2013 American Chemical Society.

SABRE

PHIP can be combined with pulse sequences such as INEPT, HMQC, etc. or with chemical reactions to enhance the detection of other sites of interest. However, a fundamental limitation of *para*-hydrogen in NMR is the requirement to generate a metal dihydride complex with inequivalent hydride ligands by addition of *para*-hydrogen to the metal. Fortunately, in favourable cases, polarization can be transferred from the hyperpolarized metal complex to another compound in an encounter complex. This process has been dubbed 'Signal Amplification By Reversible Exchange' (SABRE). In SABRE a metal-based catalyst temporarily binds both the source of hyperpolarization (*para*-H_2) and the substrate; polarization from *para*-H_2 is transferred to the substrate and is retained on dissociation from the metal. The hyperpolarized NMR spectrum of the substrate is then detected.

For example, Duckett and co-workers have demonstrated the efficacy of $[Ir(H)_2(IMes)(MeCN)_2(PPh_3)]BF_4$ **7a** and its tricyclohexyl phosphine analogue **9** as SABRE catalysts and studied the polarization transfer process using acetonitrile or pyridine as model hyperpolarization substrates. **7a** and its transformation into **7b** and **7c** were characterized by conventional ^1H and 2-D NOESY, EXSY, and $^1H,^{15}N$-HMQC NMR spectroscopy, which confirmed the structures shown in Figure 5.18a. The hydride ligands in **7c** are inequivalent, resulting in a large intensity enhancement when exposed to *para*-hydrogen. In the presence of excess pyridine in the solution, this hyperpolarization is transferred to the pyridine protons by reversible exchange of the ancilliary ligands on iridium for pyridine. This is seen in Figure 5.18c, which compares the spectra of the solution in the presence of (i) normal hydrogen; and (ii) *para*-hydrogen.

5.6 Summary

- Polarization transfer, the exchange of excitation between spins, can be used to establish connectivity through scalar coupling interactions, molecular conformation through the nuclear Overhauser effect, and to study exchange (Chapter 6).

- Polarization transfer can also be used to amplify the NMR signal from insensitive, low γ spins, S that are coupled to a sensitive, high γ spin I, the sensitivity enhancement being γ_I/γ_S when the S spin is observed directly, or $\left(\gamma_I/\gamma_S\right)^{5/2}$ when the I spin is observed (inverse, or indirect, detection).

- Hyperpolarization refers to techniques such as DNP, OPNMR, or use of *para*-hydrogen, in which highly polarized spins—i.e. spins in which the population difference between NMR energy levels has been greatly increased by some physical or chemical process—are introduced into the molecule before the NMR experiment, resulting in greatly enhanced NMR signals.

5.7 Exercises

5.1 Figure 5.19 shows the structure and the ^{119}Sn{^{1}H} refocused INEPT NMR spectrum of the polyoxometalate anion $[(MeO)SnW_5O_{18}]^{3-}$ in which there are two tungsten environments, W_{eq} and W_{ax}. The nuclear spin and abundance of the isotopes of tin and tungsten are given in Table 1.2.

  a. Explain why the INEPT experiment is preferred, rather than a simple proton decoupling experiment for observation of ^{119}Sn.

  b. Assign the couplings seen in the ^{119}Sn{^{1}H} spectrum and account for the observation that the main satellite sub-spectrum is a doublet.

  c. Predict the appearance of the ^{183}W{^{1}H} spectrum.

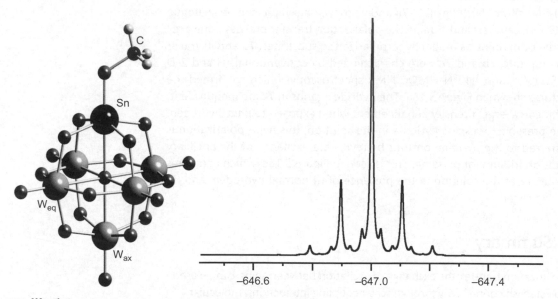

Figure 5.19 ^{119}Sn{^{1}H} refocused INEPT NMR spectrum of $[(MeO)SnW_5O_{18}]^{3-}$

Source: Reproduced with permission from B. Kandasamy et al. (2012), *Chem. Eur. J.*, **18**, 59–62. Copyright © 2011 WILEY-VCH Verlag GmbH & Co.

5.2 Figure 5.20 shows the ^{31}P,^{31}P-COSY of (NacNac)Rh(P$_5$Ph$_5$).

  a. Confirm the number of distinct phosphorus sites in the molecule.

  b. Using the information on coordination shifts given in Chapter 2, identify the resonances of P_2 and P_3.

  c. Hence assign the remaining correlations in the spectrum and attempt an assignment of the couplings in the 1-D ^{31}P{^{1}H} spectrum along the F_1 and F_2 axes.

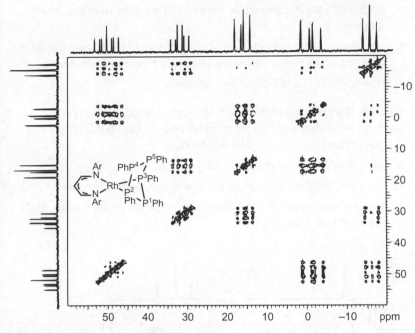

Figure 5.20 ^{31}P,^{31}P-COSY for (NacNac)Rh(P₅Ph₅)

Source: Reproduced from S. J. Geier and D. W. Stephan (2008), *Chem. Commun.*, 2779–81. Copyright 2008 The Royal Society of Chemistry.

5.3 Figure 5.21 shows the ^{1}H,^{13}C-HMQC NMR spectrum of the 1,1-dimethyl-1,5-dithiol-3-thiapentane oxorhenium complex shown. The ^{13}C spectrum has been fully assigned.

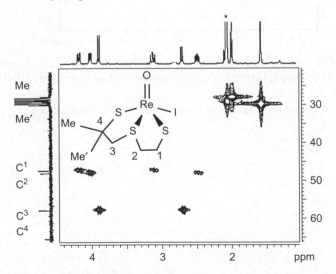

Figure 5.21 ^{1}H,^{13}C-HMQC NMR spectrum of chiral oxorhenium complex, containing 1,1-dimethyl-1,5-dithiol-3-thiapentane ligands in acetone-d_6. A residual acetone peak can be seen at 2.05 ppm.

Source: Adapted from F. De Montigny et al. (2010), *Phys. Chem. Chem. Phys.*, **12**, 8792–803. Copyright 2010 PCCP Owner Societies.

   a. Account for the occurrence of eight sets of resonances in the ^{1}NMR spectrum.

   b. Given only $^1J_{CH}$ correlations are observed, assign the ^{1}H spectrum, indicating to which carbon in the backbone each proton is bonded.

   c. How might C(1) and C(2) be distinguished?

5.4 Figure 5.22 shows part of the NOESY spectrum of the reaction product from the addition of ^{13}C-enriched MeReO$_3$ to [(bpy)PtMe$_2$] (bpy = 2,2′-bipyridine) recorded at ^{1}H = 500 MHz.

   a. Identify the resonances of the ^{13}CH$_3$ group.

   b. Use the NOESY correlations to assign the correlations observed.

   c. Hence confirm that *cis* rather than *trans* oxidative addition of MeReO$_3$ to [(bpy)PtMe$_2$] has occurred.

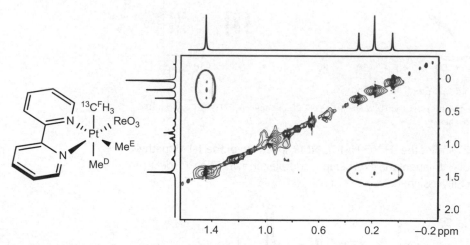

Figure 5.22 NOESY spectrum of the reaction product from addition of ^{13}C-enriched MeReO$_3$ to [(bpy)PtMe$_2$] (bpy = 2,2′-bipyridine) recorded at 500 MHz. Simulated ^{1}H NMR spectrum shown along F$_2$ to exclude impurities from the reaction. Experimental ^{1}H NMR spectrum shown in F$_1$.

Source: Adapted with permission from K. H. Pichaandi et al. (2017), *Inorg. Chem.*, **56**, 2145–52. Copyright © 2017 American Chemical Society.

6 Dynamic NMR spectroscopy

6.1 Introduction

The NMR spectrum is sensitive to a variety of molecular motions; for example, rotation about a metal-olefin bond, rotation of co-ordinated rings, conformational changes, and migration of a ligand between different sites in a molecule (Figure 6.1). These are all intramolecular processes that can be studied by NMR and are called **fluxional** processes.

NMR can also be used to study intermolecular processes such as the exchange of bound and free ligands—these are called **exchange processes**. In this chapter we will look at how NMR can be used to study fluxionality and exchange.

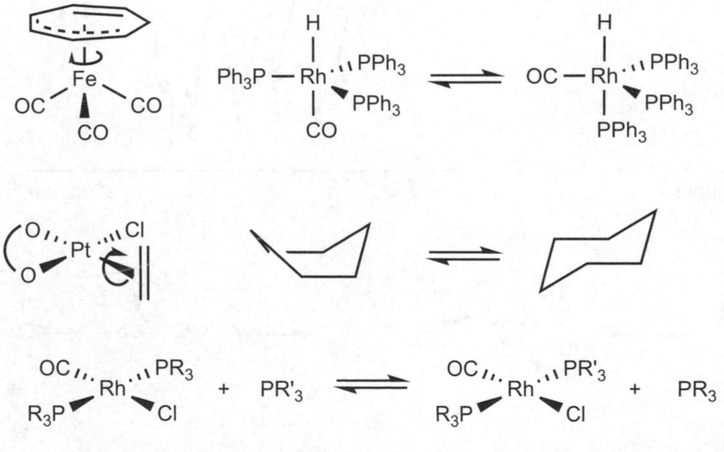

Figure 6.1 The NMR spectrum is sensitive to a variety of molecular motions, for example, rotation of co-ordinated rings, migration of a ligand between different sites in a molecule, rotation about a metal-olefin bond, conformational changes, and ligand exchange

6.2 Effect of dynamic processes on the spectra

The NMR timescale

If we look at a changing system, say a brightly coloured child's top, when the top is stationary or spinning slowly, we can see each separate segment of colour. As the top spins faster, the colours start to merge until finally we only see an average (Figure 6.2). In much the same way all resonance spectroscopies (IR, UV-Vis, NMR, etc.) can only distinguish the different sites in an exchanging system if the molecule, ligand, or whatever, remains in each site for long enough. More exactly, for separate resonances to be observed, the rate of exchange between sites must be less than $2\pi\Delta\nu$, where $\Delta\nu$ is the difference between the resonance frequencies of the sites. Typically, the NMR resonances of exchanging sites might differ by a few hertz to a few kilohertz, whereas in IR spectroscopy a difference in band position of only 1 cm^{-1} corresponds to a frequency separation of 3×10^{10} Hz! Clearly, even in very fast exchanging systems, IR will be able to distinguish each site, whereas the NMR spectrum will be sensitive to quite slow motions. We can define **NMR timescales** and IR timescales for exchange of around 10^{-7} s and 10^{-14} s, respectively; this is illustrated in Figure 6.3. Several isomers of $Co_2(CO)_8$ exist in solution, one of which adopts a bridged conformation, while at least two other conformers adopt all terminal conformations (Figure 6.3). The isomers interconvert rapidly by a fluxional process that exchanges the bridging and terminal CO ligands. By NMR we see only a single band (Figure 6.3a)—all the CO ligands look the same due to the fluxional

Figure 6.2 As a child's top starts to spin, the different segments start to blur together. When the top is spinning fast enough the separate segments can no longer be distinguished—an average is seen. In similar fashion, if nuclei move between different sites in the molecule blurring of the NMR resonances, then averaging to a single line is seen.

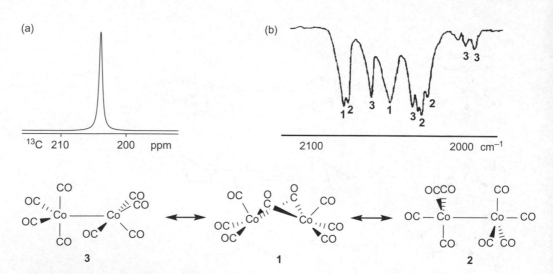

Figure 6.3 $Co_2(CO)_8$ is fluxional, interconverting between several isomers. (a) The NMR spectrum cannot distinguish even the terminal and bridging ligands; (b) the IR spectrum (terminal region only shown) of $Co_2(CO)_8$ can distinguish the absorptions of the different isomers.

Source: Adapted from R. L. Sweany et al. (1977), *Inorg. Chem.*, **16**, 415. Copyright © 1977 American Chemical Society.

process, whereas IR spectroscopy is able to distinguish not only the resonances of each isomer but also the bridging and terminal ligands (Figure 6.3b). Clearly this process must be occurring at a rate much greater than the NMR, but less than the IR, timescale.

Effect of temperature

We can alter the rate of a dynamic process by varying the temperature—at low temperatures the process will be slowed down (we are likely to be in the slow exchange regime in which we can distinguish the resonances of the different sites), whilst at high temperatures, the process will likely be fast; an averaged resonance is seen, we are in the fast exchange regime. At intermediate temperatures the lines first broaden, then coalesce into a single broad peak, which sharpens up as the **fast exchange limit** is reached. For example, $(C_5H_5)_2Fe_2(CO)_4$ exists as *cis* and *trans* isomers, which interconvert in solution. At low temperatures, separate resonances are observed in the ^1H NMR spectrum for the cyclopentadienyl ligands of the two isomers. As the temperature increases these first broaden, then coalesce, until finally a single sharp resonance is observed (Figure 6.4).

Fast and slow exchange limits

In NMR we distinguish two limiting exchange regimes:

(i) slow exchange, much slower than the difference in resonance frequencies; a separate resonance, at the characteristic resonance position of each site, is observed. The resonance is broadened by the dynamic process, the degree of broadening depending on how much time the nuclei spend at each site (the rate of site leaving); the lines become broader as the exchange rate increases. For equally populated sites the line width, $\Delta\nu$, is given by equation 6.1; k is the rate constant.

(ii) fast exchange, much faster than the difference in resonance frequencies; a single line is observed at the weighted average resonance position—in calculating the average resonance position we must remember to take account of the number of each type of site participating in the exchange/fluxional process. In the fast exchange regime, the linewidth becomes narrower as the rate of exchange increases (equation 6.2), again assuming equally populated sites. At very fast exchange rates the distinction between the sites is completely lost; the system behaves as though a single site existed. $\delta\nu$ is the separation, in Hz, of the sites in the **slow exchange limit**.

Between the fast and slow-motion regimes the resonances coalesce into a single very broad peak (equation 6.3); see for example, Figures 6.4 and 6.5.

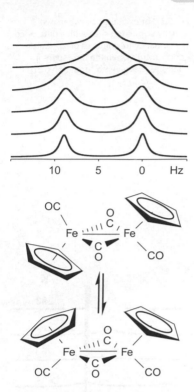

Figure 6.4 The variable temperature ^1H NMR spectra of *cis* and *trans* $(C_5H_5)_2Fe_2(CO)_4$ showing the effect of temperature on the resonances as the temperature is raised from 207 to 229 K

Source: From data in Bullitt et al. (1972), *Inorg. Chem,* **11**, 671.

Eq. 6.1 In the slow exchange limit, NMR resonances are broadened by exchange.

$$\Delta\upsilon = \frac{k}{\pi} = \frac{1}{\pi\tau}$$

Eq. 6.2 In the fast-motion regime, the linewidth narrows as the exchange rate ($\propto 1/\tau$) increases. $\delta\nu$ is the difference in resonance frequencies in the absence of exchange.

$$\Delta\upsilon = \frac{\pi(\delta\upsilon)^2}{2k} = \frac{1}{2}\pi(\delta\upsilon)^2\tau$$

Eq. 6.3 For exchange between two equally populated sites, with a difference in resonance frequencies of $\delta\nu$, coalescence occurs when the time spent in each site, τ, is given by equation 6.3.

$$\tau = \sqrt{2}\,/\,\pi(\delta\nu)$$

6.3 Determining the intimate mechanism of exchange

Concerted versus non-concerted exchange

If more than two sites, or more than two ligands, are involved in the fluxional process we can ask if the ligands are equivalenced by a single process, or does one process equivalence one pair of ligands and a second process then equivalence this pair with the third ligand, etc., i.e. do we have a single, *concerted process* or two *non-concerted processes*? For example, the *cis–trans* isomerism in $(C_5H_5)_2Fe_2(CO)_4$ might involve localized rotation about one iron centre with no concomitant equivalencing of the bridging and terminal CO ligands or via rotation about the iron–iron bond in an all-terminal isomer, in which case the fluxional process must simultaneously equivalence all the CO ligands. Similarly, if ligand exchange is involved in a fluxional process, we might ask is the exchange process associative or dissociative? We can answer such questions by analysis of the variable temperature (VT) NMR spectra to obtain the rate constants for the observed dynamic processes.

Consider exchange of the pyridylphosphine ligands in $[Pd(PPh_2py)_3]^{2+}$ (Figures 6.5 and 6.6). Separate resonances are observed in the $^{31}P\{^1H\}$ NMR spectrum for each phosphorus at low temperature, but at ambient temperature a single resonance is seen, indicating a fluxional system in which all the coordination sites are rapidly equivalenced. Spectra recorded at intermediate temperatures show coalescence of P^B and P^C at 233 K followed by coalescence with the resonance of P^A at 283 K. A simple analysis might consider this as evidence for two separate processes, but this approach is incorrect. The rate of exchange required for coalescence depends on the separation of the resonances, $\delta\nu$, in the frozen-out spectrum (equation 6.3). This means that the widely separated pair P^B/P^C must require a faster rate of exchange to coalesce than the rate required for coalescence of P^A and P^B which have much more similar chemical shifts, i.e. we expect different coalescence temperatures irrespective of whether there is one, concerted, or two, non-concerted, processes occurring. N.B. To distinguish a concerted from a non-concerted process we must determine the rate constant for the exchange(s) by simulating the VT-NMR spectra. Simulation of the NMR lineshapes with temperature allows us to extract not only the rate constants but also the thermodynamic parameters $\Delta G^\ddagger$, $\Delta H^\ddagger$, and $\Delta S^\ddagger$ for the exchange. In this example, these are different for exchange of P^B with P^C versus the P^B/P^C with P^A, confirming a non-concerted mechanism. Exchange of P^B with P^C is essentially a flip-flop exchange via a square-based pyramidal intermediate (Figure 6.6 (top)) while exchange of P^B/P^C with P^A requires a sequence of Berry pseudo rotations of the square-based pyramidal intermediate formed on coordination of the N of ligand P^A (Figure 6.6 (bottom)), explaining the higher E_{act} of the latter process.

The spectra also allow us to establish that the dynamic processes are intramolecular, since the resonance of free Ph_2Ppy remains sharp below the coalescence temperature of both processes. Since the amount of Ph_2Ppy present is small and its resonance remains sharp, it cannot be involved in either fluxional process due to the absence of differential broadening—see below.

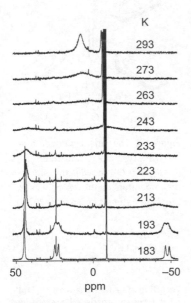

K
293
273
263
243
233
223
213
193
183

50   0   −50
ppm

Figure 6.5 Variable temperature $^{31}P\{^1H\}$ NMR spectra indicate non-concerted intramolecular fluxional processes equivalence the coordination sites in $[Pd(PPh_2py_3)]^+$. The resonances of free PPh_2py (δ ca. −10 ppm) and of the impurity (δ ca. 25 ppm) remain sharp below 243 K, indicating these species are not involved in the fluxional processes. py = pyridine.

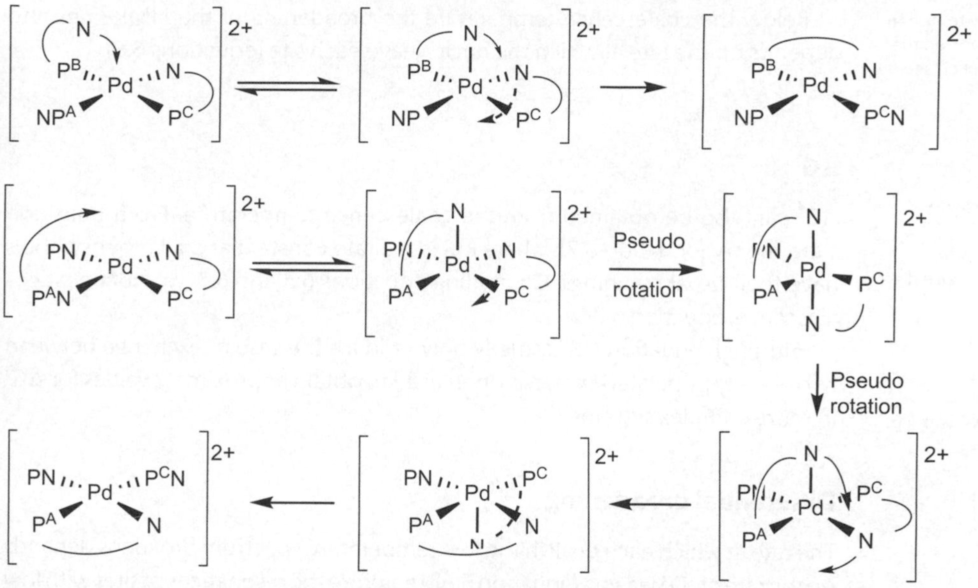

Figure 6.6 Two, non-concerted, dynamic processes equivalence the pyridyl phosphine ligands in [Pd(PPh$_2$py$_3$)]$^+$. A low-temperature process involves associative intramolecular exchange of the dangling pyridyl N of Ph$_2$P^{B}py with the chelating N of Ph$_2$P^{C}py (top). Exchange with Ph$_2$P^{A}py requires a multistep pathway involving several Berry pseudo rotations accounting for the higher activation barrier (bottom).

Source: From data in Liu et al. (2010), *Dalton Trans.*, **39**, 7921–35.

6.4 Determination of the thermodynamic and kinetic parameters of exchange

There are several strategies available to determine the thermodynamic and kinetic parameters of a dynamic process from the NMR spectrum, including determination of the coalescence temperature, lineshape analysis, saturation transfer, and 2–D methods.

The rate constant

Consider the equilibrium shown in equation 6.4 in which exchange is occurring between two sites of unequal population, for example, a co-ordinated ligand with free ligand in solution (Figure 6.1). We can obtain the forward and backward rates of the reaction from the broadening of the NMR resonances of the two types of site.

Suppose the two sites have unequal populations p_A and p_B, where $p_A + p_B = 1$, then at equilibrium, $p_A k_A = p_B k_B$ and we need to modify equation 6.2 to account for the different populations of the two sites. The linewidth above the coalescence point, is now given by equation 6.5.

Eq. 6.4 The forward and backward rates of the exchange between two sites of unequal population can be obtained from the broadening of the NMR resonances of the two sites.

$$A \underset{k_B}{\overset{k_A}{\rightleftharpoons}} B$$

Eq. 6.5 The NMR linewidth for exchange between two sites of unequal population above coalescence is given by equation 6.5.

$$\Delta v = \frac{4\pi p_A p_B (\delta v)^2}{k_A + k_B}$$

Eq. 6.6 Below the coalescence temperature the NMR linewidth depends on the
rate at which the nuclei leave each site,
equations 6.6.

$$\Delta v_A = \frac{k_A}{\pi} = \frac{1}{\pi\tau_A}, \quad \Delta v_B = \frac{k_B}{\pi} = \frac{1}{\pi\tau_B}$$

Eq. 6.7 From transition state theory
we know that $\Delta G^\ddagger$ is given by
equation 6.7, where k_{rxn} is the rate
constant and all other symbols have their
usual meanings.

$$k_{rxn} = \frac{kT}{h}e^{\left(-\Delta G^\ddagger / RT\right)} = \frac{1}{\tau}$$

Eq. 6.8 Combining equations 6.3 and 6.7
yields $\Delta G^\ddagger$ at coalescence.

$$\frac{\Delta G^\ddagger}{RT_c} = \ln\left(\frac{\sqrt{2}kT_c}{\pi(\delta v)h}\right)$$

Below the coalescence temperature the broadening of the NMR lines also depends on the rate at which the nuclei leave each site (equations 6.6).

$\Delta G^\ddagger$

$\Delta G^\ddagger$ can also be obtained from the coalescence temperature. From transition state theory (equation 6.7): where k_{rxn} is the rate constant and all other symbols have their usual meanings. Combining equations 6.3 and 6.7, we obtain, at coalescence, equation 6.8.

Although equation 6.8 is strictly only valid for the case of exchange between two equally populated sites, it is often used to obtain approximate values for $\Delta G^\ddagger$ in more complex systems.

Differential broadening

The rate at which each peak in the low temperature spectrum broadens depends on the rate of site leaving (equation 6.6), therefore the resonances of sites with low populations broaden first. This phenomenon is called differential broadening.

Crown ethers and cryptands form an important class of ligand that can show high selectivity for particular metal ions based on size, and find application in metal ion extraction, catalysis, medicine, biomimetic research, as receptors and electrides, and as models for antibiotics of the valinomycin and nigericin families. The co-ordination–deco-ordination equilibrium of a metal ion by a cryptand ligand can be considered as a two-site exchange of free and complexed metal ions. Figure 6.7 shows the effect of differential broadening on the resonances of free and complexed Li^+ ions of the relative concentrations of the cryptand and metal ion. At low metal:cryptand ratio (Figure 6.7a), the ^7Li NMR resonance of the free metal ion broadens more quickly than the resonance of the bound metal ion as the temperature is increased. This reflects the difference in concentration and hence the different rates of site leaving, of the two species. For every Li^+ that is released from the cryptand complex, a Li^+ ion from solution coordinates; since there are more coordinated Li^+ than free Li^+ ions, the rate of site leaving for each free ion must be higher than the rate of site leaving of the coordinated Li^+ ions. At higher lithium:cryptand ratios, the situation is reversed and the resonance of the coordinated Li^+ ions broadens more quickly. A detailed analysis of the data revealed that an associative pathway operates at low temperature, whereas both associative and dissociative pathways exist at higher temperature (Figure 6.8).

Multisite dynamic processes

Lineshape analysis can also be used to study multisite exchange. For example, exchange of all the carbonyl ligands in $Rh_4(CO)_{12}$ is proposed to occur by a fluxional process in which the bridges open and then reform about any one of the four tetrahedral faces—the 'Cotton' merry-go-round (Figure 6.9a). Lineshape analysis of the variable temperature; ^{13}C NMR spectra of the cluster has confirmed this proposal and allowed the free enthalpy of activation, $\Delta H^\ddagger = 42.8 \pm 0.4$ kJmol^{-1} at 298 K, to be determined.

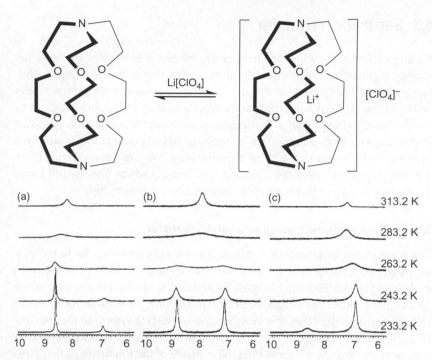

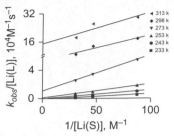

Figure 6.8 Plots of k_{obs}/[Li(L)] vs 1/[Li(S)] indicate that both dissociative and associative pathways are followed. At low temperature the plots exhibit no meaningful intercept, an associative pathway is dominant, whereas at high temperature large intercepts are seen, indicating a significant dissociative contribution to the exchange at high temperature.

Source: Adapted from E. M. Pasgreta et al. (2007), *Eur. J. Inorg. Chem.*, 3067–76. Copyright © WILEY-VCH Verlag GmbH, 69451 Weinheim, 2007.

Figure 6.7 The variable temperature ^{7}Li NMR spectra of mixtures of Li[ClO$_4$] and a cryptand ligand show differential broadening of the resonances since the rate of site leaving depends on the relative concentrations of free and bound Li$^+$ ions

Source: Adapted from E. M. Pasgreta et al. (2007), *Eur. J. Inorg. Chem*, 3067–76. Copyright © WILEY-VCH Verlag GmbH, 69451 Weinheim, 2007.

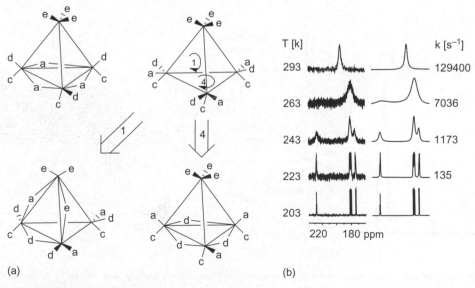

Figure 6.9 (b) Measured and calculated ^{13}C NMR spectra of Rh$_4$(CO)$_{12}$ at different temperatures and (a) the proposed fluxional process. Simulation of the NMR lineshapes allowed the rate constants as a function of temperature to be determined. A single fluxional process occurs: the bridging COs become terminal and symmetry-equivalent merry-go-rounds about any triangular face of the tetrahedron then result in complete, intramolecular scrambling of the CO ligands about a rigid metal core.

Source: Adapted with permission from K. Besancon et al. (1998), *Inorg. Chem.*, **37**, 5634. Copyright ©1998 American Chemical Society.

6.5 Saturation transfer

An alternative strategy for obtaining rate constants is to work in the slow exchange regime and selectively excite the nuclei at one of the exchanging sites. Since NMR relaxation is a slow process (Chapter 4, section 4.4) the excited nuclei will carry the excitation with them as they move from one site in the molecule to another, and so the excitation will appear at each site the excited nucleus visits during the exchange process. This is much like heating up a cup of water then pouring the hot water into another cup—the water takes its heat with it. If there are several possible alternative cups, we can find out which one the hot water has been poured into by monitoring the temperature of each cup.

Monitoring exchange using saturation transfer

There are several experimental methods that use saturation transfer to monitor exchange. The conceptually simplest method is to saturate one resonance in the exchanging system (Figure 6.10a (middle)) and then wait a little to allow exchange to take place before recording a spectrum. The saturation will move from the irradiated peak to all the other sites visited in the exchanging system so the intensity of the resonances of sites to which the saturated spins have migrated will reduce (Figure 6.10a, top). At the same time the intensity of the irradiated peak will recover as unsaturated spins move from the other sites to the site that was saturated. The exchanging system can thus easily be identified by the change in the intensities of the various resonances with those in an 'ordinary' NMR spectrum of the sample (Figure 6.10a (bottom)). A similar result is obtained if a selective excitation pulse is used to invert one resonance in an exchanging system (Figure 6.10b).

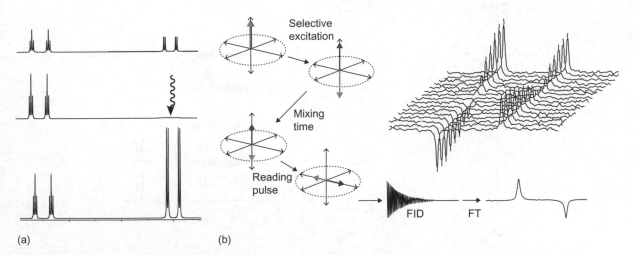

(a)   (b)

Figure 6.10 Exchange pathways can be mapped out by (a) saturating; or (b) selectively exciting, one resonance in the exchanging system and monitoring the intensity of the other resonances. These will become less intense as saturation is carried by exchange around the sites involved in the exchange.

Source: Adapted from I. Banyai (2018), *New J. Chem.*, **42**, 7569. Copyright © 2018 The Royal Society of Chemistry.

Measuring exchange rates using saturation transfer

Saturation transfer methods follow the movement of excitation around the exchanging system with time, so can be used to determine the rates of intra- and inter-molecular exchange. The rate at which the excitation appears at other sites in the dynamic system can be measured by inserting a variable time delay (the mixing time, Figure 6.10b), during which exchange occurs between the excited and the other sites involved in exchange, before measuring the spectrum. For example, saturation transfer has been used to study the fluxional processes and intermolecular ligand exchange with free PPh_3 in Wilkinson's catalyst. A DANTE pulse sequence was used to invert one branch of the *trans*-PPh_3 resonance and a series of spectra with different mixing times recorded (Figure 6.11).

Analysis of the NMR data allowed the rate constants for fluxional exchange of $^aPPh_3$ with $^bPPh_3$ and for associative and dissociative exchange with free PPh_3 and the potential reaction pathways to be determined. Subsequent NMR and density functional theory (DFT) studies extended the early work of Brown's and Heaton's groups to a range of analogues of Wilkinson's catalyst $[Rh(PPh_3)_3X]$, (X = H, Me, Ph, CF_3, Cl) and confirmed their conclusions that the fluxional process in $[Rh(Ph_3P)_3Cl]$ occurs *via* a distorted tetrahedral intermediate, and intermolecular exchange with free PPh_3 is dissociative (Figure 6.12). Thus, **intermolecular exchange** of free and bound PPh_3 in $[Rh(PPh_3)_3Cl]$ is considerably slower than equivalencing of $^aPPh_3$ with $^bPPh_3$, implying the latter is an **intramolecular exchange** process.

Saturation transfer by spin diffusion

Saturation can also be transferred by spin diffusion in a host–guest encounter complex. Such experiments are widely used to study ligand binding to proteins,

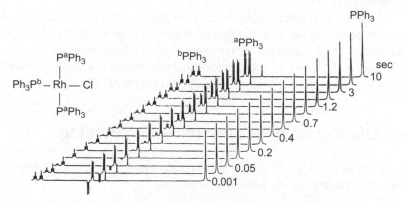

Figure 6.11 $^{31}P\{^1H\}$ NMR spectra showing saturation transfer between the *cis* and *trans* coordinated PPh_3 ligands in Wilkinson's catalyst. The downfield branch of the resonance of $^aPPh_3$ was selectively inverted using the DANTE pulse sequence. From t = 0.05 to 5 s a significant loss of intensity of the downfield branch of $^bPPh_3$ due to interchange of $^aPPh_3$ and $^bPPh_3$ is seen.

Source: Reproduced from J. M. Brown et al. (1987), *J. Chem. Soc. Perkin*, **2**, 1589. Copyright © 1987 The Royal Society of Chemistry.

Metal-organic frameworks are porous coordination polymers in which poly dentate bridging ligands link/cross-link the metal centres to form an extended molecular solid. They have potential applications as, e.g., gas storage, drug delivery, and reaction media, and as catalysts.

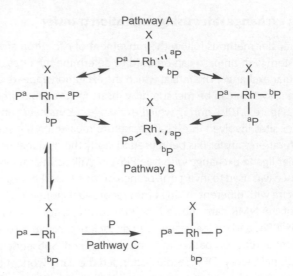

Figure 6.12 Intramolecular exchange pathways in $[Rh(PPh_3)_3X]$ determined by NMR and DFT studies. Pathway A is favoured for strongly electron donating X. Intermolecular exchange occurs via a dissociative pathway involving only P^a.

Source: From data in J. Goodman, et al. (2010), *J. Amer. Chem. Soc.*, **132**, 12013.

but can also be applied to study host–guest systems. In these experiments a resonance of the host is saturated, a delay inserted to allow transfer of the excitation to the guest through non-covalent interactions in the encounter complex, and a spectrum then recorded. The experiment is repeated with the saturating radiation at a frequency far from any resonances, and the difference spectrum calculated. If the guest is interacting with the host, the difference spectrum will show the resonances of the guest.

This can be seen in the saturation transfer difference (STD) spectrum of naphthaldehyde interacting with the metal organic framework (MOF) Al–ITQ–HB. Saturation was applied to the methyl signal of Al–ITQ–HB (0.9 ppm) and far from resonance (−150 ppm). All of the naphthaldehyde resonances appear in the STD spectrum, demonstrating that naphthaldehyde binds to Al–ITQ–HB (Figure 6.13).

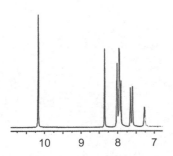

Figure 6.13 STD spectrum of naphthaldehyde/Al–ITQ–HB. The resonances of naphthaldehyde are not excited directly; rather polarization has been transferred from the Al–ITQ–HB spins to those of naphthaldehyde in a host–guest encounter complex; cf. the cross-polarization experiment in solid-state NMR.

Source: Reproduced from P. García-García, et al. (2016), *Nature Commun.*, **7**, 10835. This work is licensed under a Creative Commons Attribution 4.0 International License.

6.6 Effect of exchange on spin–spin coupling

The effect of exchange on coupling differs between a fluxional, intramolecular process and an intermolecular exchange process.

Averaging of the coupling constant in an intramolecular process

In a fluxional process the ligand/migrating group is bound at all times to the molecule and hence the coupling is retained; the observed coupling constant is the weighted average of the individual couplings and the coupling pattern is that expected, assuming all the coupling partner nuclei are visited.

Consider for example the $^{31}P\{^1H\}$ NMR spectrum of $[Rh(PPh_3)_3]ClO_4$ in dichloromethane solution (Figure 6.14). At low temperature this shows two resonances, a doublet of triplets and a doublet of doublets in the ratio 1:2 due to the unique ($^1J(RhP)$ = 244 Hz) and *trans* phosphines ($^1J(RhP)$ = 134 Hz), respectively. As the temperature is increased, both resonances broaden and then coalesce to form a doublet ($^1J(RhP)$ = 172 Hz, the weighted average Rh–P coupling constant) at 38.1 ppm, the weighted average chemical shift. Note that the resonances of the unique phosphine broaden more quickly than those of the pair of *trans* phosphines, reflecting differential broadening due to the unequal population of the two sites. The phosphorus ligands are now equivalent so no longer couple to each other. The preservation of $^1J(RhP)$ at high temperatures indicates that the Rh–P bonds are conserved in the fluxional process, consistent with an *intramolecular* fluxional process that exchanges the inequivalent triphenylphosphines.

Exchange decoupling in an intermolecular process

In an intermolecular exchange process the bond between the exchanging group and the rest of the molecule breaks. Since coupling is transmitted through the bonds, decoupling will occur when the exchange rate—the rate of bond breaking—is greater than the coupling constant. This is known as exchange decoupling; the well-known absence of coupling between the OH proton in methanol and the other protons in the molecule is not due to the intervening oxygen atom but to rapid intermolecular exchange of this proton catalysed by traces of acid or base in the sample. If the alcohol is rigorously purified the expected couplings are seen.

Exchange decoupling is seen in the ^{205}Tl NMR spectra of aqueous solutions of $[Tl(^{13}CN)_4]^-$ containing excess $H^{13}CN$. On increasing the pH deprotonation of HCN occurs, increasing the concentration of free cyanide ions, so the rate of intermolecular exchange of $^{13}CN^-$ increases and the ^{205}Tl and ^{13}C spins are exchange decoupled; $J(^{205}Tl^{13}C)$ is lost and the quintet collapses to a singlet (Figure 6.15). Consideration of the pseudo first-order rate constants as a function of pH indicated an associative mechanism for the exchange.

6.7 2-D methods for studying exchange

So far we have described variable temperature and 1-D saturation transfer methods for studying exchange. In those experiments we looked at relatively simple systems where the identification of the dynamic process was straightforward. However, in more complex systems in which there are several, possibly interdependent exchange pathways, these 1-D methods may give ambiguous results or require many separate 1-D saturation transfer experiments exciting each site in turn to map out the individual ligand movements. It would clearly be

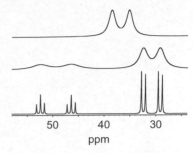

Figure 6.14 The variable temperature $^{31}P\{^1H\}$ NMR spectra of $[Rh(PPh_3)_3]$ ClO_4 showing the effect on spin–spin coupling of intramolecular exchange. At low temperature (lower spectrum) P–P coupling and discrete P–Rh couplings are seen, whereas at high temperature (upper spectrum) an averaged coupling to rhodium is seen.

Source: From data in A. R. Siedle et al. (1988), *Inorg. Chem.*, **27**, 2473.

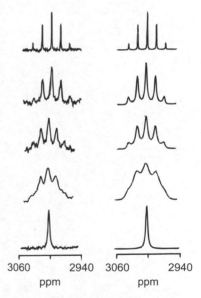

Figure 6.15 ^{205}Tl NMR spectra of aqueous solutions of $[Tl(^{13}CN)_4]^-$ containing excess $^{13}CN^-$. $J(^{205}Tl^{13}C)$ is lost as the exchange rate between coordinated and free $^{13}CN^-$ increases.

Source: Reproduced from I. Bányai et al. (2001), *Eur. J. Inorg. Chem.*, 1709. Copyright © WILEY-VCH Verlag GmbH, 69451 Weinheim, 2001.

advantageous if we could excite all sites simultaneously in a saturation transfer experiment. In effect this is what the 2-D EXSY (EXchange SpectroscopY) experiment achieves.

Exchange spectroscopy

In the EXSY experiment we first excite all the nuclear spins in the sample (cf. the 1-D saturation transfer in which we selectively excited the nuclear spin at a particular site), and then allow a mixing delay for exchange to occur before measuring the free induction decay (FID). The experiment is repeated using a series of mixing delays to create a second time dimension—see Chapter 5—that contains information about the exchange processes occurring. Fourier transformation of the FIDs in both time domains generates a 2-D map in which (in addition to the diagonal peaks) peaks appear linking sites between which exchange is occurring.

For example, the phosphorus containing ligand in $[Pd(ArF)_2(SPPy_2Ph-S,N)]$ is N,S-bonded, making the two fluorinated aryls inequivalent in a static structure. Through space, dipolar coupling allows assignment of the fluorine signals. The ^{19}F EXSY spectrum (Figure 6.16), reveals a pairwise fluxional process that equivalences the ArF ligands. The fluxional process is ascribed to a *turnstile* mechanism in which co-ordination–deco-ordination of the free and bound pyridines occurs: the incoming and leaving ligands approach and leave the same face of the square–planar complex, with the incoming pyridine displacing the sulfur, which migrates to the site of the leaving pyridine.

Multisite exchange

In most of the examples above there were two sites involved in the exchange process; the cyclopentadienyl rings in $(C_5H_5)_2Fe_2(CO)_4$ were either mutually *cis* or *trans*, the two ArF ligands in the Pd complex in 'Exchange spectroscopy' above, or free and co-ordinated metal ions or ligands were exchanged. However, in many systems there are more than two sites involved in the fluxional process and it can be difficult, using variable temperature methods, to determine exactly

Analysis of EXSY spectra follows a similar procedure to analysis of COSY and NOESY spectra; look for 'squares' that link off-diagonal with diagonal peaks—see the dotted lines in Figure 6.17.

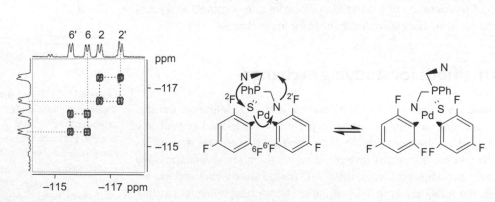

Figure 6.16 Part of the ^{19}F EXSY spectrum of $[Pd(ArF)_2(SPPy_2Ph–S,N)]$ showing pairwise exchange of ^2F with $^{2'}F$ and ^6F with $^{6'}F$

Source: Adapted from P. Espinet et al. (2008), *Coord. Chem. Rev.*, **252**, 2180–208. Copyright © 2008 Elsevier B.V.

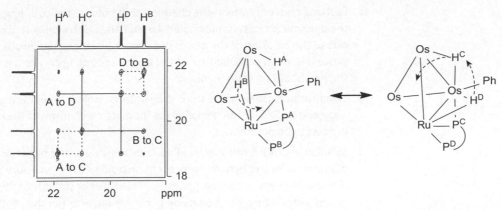

Figure 6.17 EXSY ^{1}H NMR spectrum of [RuOs$_3$(μ–H)$_2$(CO)$_{10}$[μ,κ^2:κ^1–PPh(CH$_2$)$_3$PPh$_2$](Ph)]

Sources: Adapted from data in Y. L. K. Tan and W. K. Leong (2007), *J. Organomet.Chem.*, **692**, 2253–69 and Y. L. K. Tan (2007), Ph.D. thesis, National University of Singapore, http://scholarbank.nus.edu.sg/handle/10635/16255.

which ligand moves where. EXSY is particularly useful in studying dynamic processes in complex systems involving multisite exchanges in which the exact atom movements cannot readily be obtained from conventional variable temperature measurements.

Virtual exchange

Virtual exchange occurs when a symmetry transformation, as a result of movement of some other ligand in the compound, equivalences two otherwise distinct sites. Virtual exchange is thus likely to occur in systems where multisite exchange is possible.

The EXSY ^{1}H and ^{31}P{^{1}H} NMR spectra of [RuOs$_3$(μ-H)$_2$(CO)$_{10}$[μ,κ^2:κ^1-PPh(CH$_2$)$_3$PPh$_2$](Ph)] (Figure 6.17) reveal a fluxional process that interconverts H^A and H^B via a stepwise merry-go-round about the apical-Os-Ru(μ-P)Os face of the tetrahedron. In this way H^B becomes H^D and H^A becomes H^C (dashed squares in Figure 6.17). Then H^D moves to the H^C site (previously occupied by H^A), while H^C (previously H^A) moves to occupy the site originally occupied by H^B. The lower intensity of the correlation peaks for H^A–H^B exchange is consistent with this stepwise process. Although the ^{31}P{^{1}H} EXSY spectrum (not shown) reveals 'exchange' of P^A with P^C and P^B with P^D, the Os$_3$Ru(PP) core remains static throughout the fluxional process, providing an example of virtual exchange.

6.8 Summary

- Intra- and intermolecular dynamic processes affect the appearance of NMR spectra. When the dynamic process is slow, separate resonances are observed for each site. When the process is fast, a resonance at the weighted average chemical shift is seen.

- Exchange not only affects the chemical shift but also the couplings seen. If the dynamic process is intermolecular exchange decoupling is observed due to the breaking of the bonds between the exchanging groups. If the process is intramolecular a weighted average coupling constant is seen to all the neighbouring spins involved.

- Dynamic processes can be studied by both variable temperature spectroscopy and by EXSY NMR, allowing the intimate mechanism of the dynamic process to be determined.

- Simulation of VT NMR spectra allows rate constants, and hence activation parameters for the dynamic process to be obtained. Where more than one dynamic process occurs, comparison of the rate constants, not the coalescence temperature, must be made to determine whether the processes are concerted or independent.

6.9 Exercises

1. Figure 6.18 shows the variable temperature ^{31}P NMR spectra of $[Ni(L)_2]$, $L = SeSPPh_2$. Suggest a dynamic process that accounts for the high- and low-temperature spectra.

2. Figure 6.19 shows the ^{19}F NMR spectra of $AgBF_4$ in acetone recorded at low, intermediate, and high temperature. Explain why two chemical shifts, with different coupling patterns and intensities, are expected at low temperature. Suggest an explanation for the collapse of the resonances at high T to a singlet, at a shift closer to that of the low T quartet than the low T septet. Suggest two possible reasons why the coupling in the septet is lost before that in the quartet.

Hint: What are the isotopologues present?

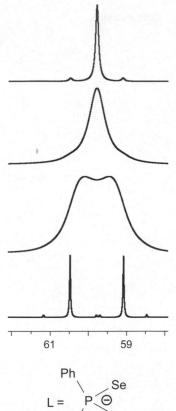

Figure 6.18 Simulated VT ^{31}P NMR spectra of $[Ni(SeSPPh_2)_2]$

Source: From data in A. V. Artem'eva et al. (2014), *J. Organomet. Chem.*, **768**, 151–6.

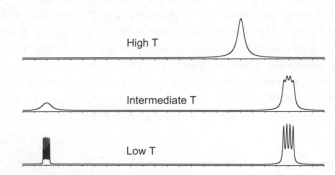

Figure 6.19 Simulated VT ^{19}F NMR spectra of $AgBF_4$

Source: R. J. Gillespie (1968), *Can. J. Chem.*, **46**, 1601–10.

3. Account fully for the temperature dependence of the ^{13}C NMR spectra of [Rh(PBu$_3$)$_2$(CO)Cl] in the presence of traces of dissolved CO (Figure 6.20).

4. Figure 6.21a shows the variable temperature $^{31}P\{^1H\}$NMR spectrum of a mixture of [PPh$_3$Cl]Br and [PPh$_3$Cl]Br. Account for the observation of separate resonances at −60 °C but a single resonance at +20 °C. Figure 6.21b also shows the EXSY spectrum of a mixture **1**−X. Identify which species participate in exchange and assign the exchanging pairs. By considering the chiral information, decide if exchange occurs with retention or inversion of stereochemistry at phosphorus.

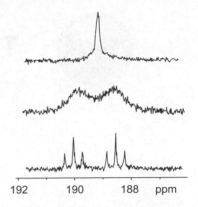

Figure 6.20 ^{13}C VT NMR spectra of [Rh(PBu$_3$)$_2$(CO)Cl] in the presence of traces of CO

Source: J. A. Iggo, unpublished.

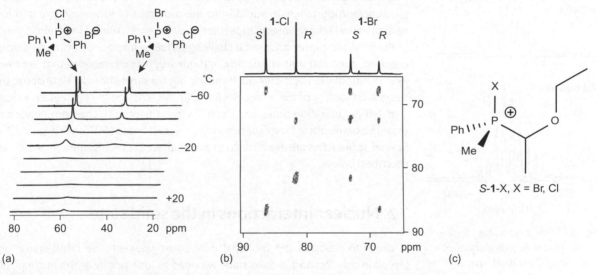

(a)   (b)   (c)

Figure 6.21 VT $^{31}P\{^1H\}$ NMR spectrum of [PPh$_3$Cl]Br and [PPh$_3$Br]Cl (left); the $^{31}P\{^1H\}$ EXSY NMR spectrum of a mixture of *R*− and *S*−[PPhMeRCl]$^+$ and [PPhMeRBr]$^+$

Source: Reproduced with permission from K. Nikitin et al. (2018), *Angew. Chem. Int. Ed.*, **57**, 1480–4. Copyright © WILEY-VCH Verlag GmbH & Co. KGaA, Weinheim.

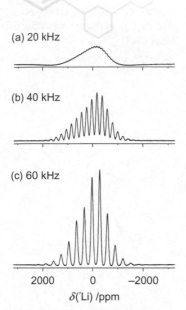

7 The solid state

N/A

(a) 20 kHz

(b) 40 kHz

(c) 60 kHz

2000   0   −2000

$\delta(^7Li)$ /ppm

Figure 7.1 The effect of magic angle spinning on the ^{7}Li NMR spectrum of LiFe$_{0.5}$Mn$_{0.5}$PO$_4$. At 20 kHz spinning rate (a) a broad, featureless resonance is seen. (b) Spinning at 40 kHz reduces the broadening effect of interactions between the ^{7}Li nuclei and the surrounding Fe(II) and Mn(II) ions, giving sharper lines. (c) At 60 kHz MAS baseline resolution of the spinning sidebands is seen.

Source: Reproduced with permission from A. J. Pell and G. Pintacuda (2015), *Prog. Nucl. Mag. Reson. Spectr.*, **84–85**, 33–72. Copyright © 2015 Elsevier Ltd. All rights reserved.

Eq. 7.1 The NMR Hamiltonian is divided into interactions with the external magnetic field $\hat{H}_{ext}$ and interactions between the spins $\hat{H}_{int}$.

$$\hat{H} = \hat{H}_{ext} + \hat{H}_{int}$$

7.1 Introduction

Although NMR spectroscopy in solution continues to be the most widely used magnetic resonance spectroscopic technique, NMR in the solid state is rapidly growing in importance. In addition to the problems of low sensitivity and low natural abundance for many important nuclides encountered in solution, NMR in the solid state faces additional challenges due to interactions such as dipolar coupling, chemical shift anisotropy, and quadrupolar interactions that are averaged out in solution by molecular tumbling but remain in the solid state due to the restricted mobility of the molecules/ions. Restricted mobility also results in long spin–lattice relaxation times and short T_2s (see Chapter 4). The combination of these factors results in long experimental times and broad NMR lines (Figure 7.1a). Several approaches are available to deal with these problems, some of which are described below.

7.2 Nuclear interactions in the solid state

In order to describe the principal differences between the NMR signal observed in solution and in solid state, we need to look briefly at the nuclear spin Hamiltonian. First, we divide the total Hamiltonian into $\hat{H}_{ext}$, which describes interactions with the external magnetic field, and $\hat{H}_{int}$, dealing with interactions between spins in the system (equation 7.1).

$\hat{H}_{ext}$ describes the interaction with the spectrometer field, $\textbf{\textit{B}}_0$ (the Zeeman interaction, see Chapter 1) and with the pulses used to excite the nuclear spins, $\textbf{\textit{B}}_1$ (see Chapter 4). In this chapter we concern ourselves with the interactions in $\hat{H}_{int}$ (equation 7.2).

$$\hat{H}_{int} = \hat{H}^{CS} + \hat{H}^{SR} + \hat{H}^{D} + \hat{H}^{J} + \hat{H}^{Q} \qquad \text{(Eq. 7.2)}$$

$\hat{H}^{CS}$ describes the chemical shift, $\hat{H}^{D}$ describes dipolar interactions between spins which are important in the solid state, while $\hat{H}^{J}$ describes scalar coupling with which we are familiar from solution NMR. $\hat{H}^{SR}$ describes spin rotation and $\hat{H}^{Q}$ quadrupolar coupling (for $\textbf{\textit{I}} > ½$ nuclei only). Chemical shift anisotropy and dipolar interactions are discussed in the following sections, but spin rotation, and quadrupolar coupling (for $I > \frac{1}{2}$ nuclei only), while important, are beyond the scope of this book.

$\hat{H}^{CS}$, $\hat{H}^D$, and $\hat{H}^Q$ all depend on the orientation of the individual molecules/ions with respect to $\boldsymbol{B_0}$, i.e. they will be anisotropic (Figure 7.2(top)). The term describing this orientational dependence contains the factor $(3\cos^2\theta - 1)$; in solution, rapid tumbling (Brownian motion) averages $(3\cos^2\theta - 1)$ to zero and high-resolution spectra (linewidths typically less than 0.1 Hz for spin ½ nuclei) are obtained. In the solid state, however, molecular motion is restricted so cannot average out these interactions, but we can trick the system and reduce these line-broadening interactions by spinning the sample at the 'Magic Angle' (Figure 7.3).

Magic angle spinning (MAS)

$3\cos^2\theta-1$ equals zero when θ is 54.74°. Mechanical rotation of the sample at this angle reduces the spatial term of the Hamiltonian. If the spinning rate is fast enough, then the orientation dependence of the chemical shift and dipolar interactions can be almost completely removed and sharp spectra obtained (Figures 7.1, 7.2). Quadrupolar effects, however, are only partially removed by magic angle spinning (MAS) due to the second-order quadrupolar effect. Although the static lineshape contains much valuable information, the overwhelming majority of contemporary **solid-state NMR** experiments are done using MAS.

7.3 Chemical shift anisotropy

Consider the tetrahedral molecule triphenylphosphine oxide (Figure 7.2); clearly the electrons are not isotropically distributed around the phosphorus. In a powder sample in which the powder grains will be randomly oriented, each Ph_3PO molecule will be in a different, fixed orientation with respect to the spectrometer magnetic field. Different shielding—i.e. a different chemical shift—is expected for each orientation. If we record the $^{31}P\{^1H\}$ NMR spectrum of a powder sample of Ph_3PO, all possible orientations will be present, and we get the spectrum shown in Figure 7.2a, in which the lineshape encompasses all these orientation-dependent shifts. Spinning the sample at the magic angle reduces the chemical shift anisotropy (CSA) and sharp lines are seen. The appearance of the MAS spectrum depends on the spinning rate, which determines how effectively the $3\cos^2\theta - 1$ term is averaged out (Figures 7.2b, c).

Spinning sidebands

If the spinning rate is much faster than the width of the static lineshape in Hz, a sharp resonance is seen at the isotropic chemical shift, σ_{iso}. At slower spinning rates, the orientation-dependent term is not completely removed, and spinning side bands are seen. This effect can be seen in the solid state $^{31}P\{^1H\}$ NMR spectra of Ph_3PO (Figure 7.2), and of $NH_4H_2PO_4$ (Figure 7.4). A further advantage of MAS is that the spectral intensity is focused into a few sharp lines, greatly improving signal to noise. The sideband envelope can be simulated, and the principal components of the chemical shielding tensor obtained.

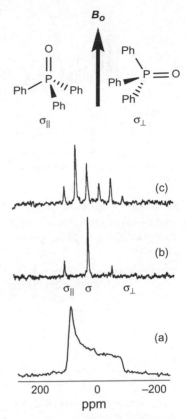

Figure 7.2 The interaction between the nuclear spins in a compound and the spectrometer field will depend on the angle between them. (a) The solid-state ^{31}P NMR spectrum of powdered Ph_3PO showing the locations of $\sigma_{\parallel},\sigma_{\perp}$, and σ_{iso}; (b), (c) showing the effect of MAS.

Source: Adapted from R. E. Wasylishen et al. (1982), Ann. Rep. NMR Spectroscopy, 1. G. A. Webb ed., Copyright © 1982 Academic Press, London.

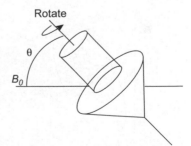

Figure 7.3 Schematic representation of the MAS experiment. Spinning the sample at an angle 54.7° to the magnetic field averages the orientational scaling factor $(3\cos^2\theta-1)$ to zero.

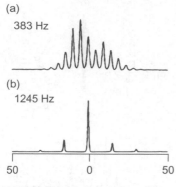

(a)

383 Hz

(b)

1245 Hz

50   0   50

Figure 7.4 ^{31}P MAS NMR spectra of a solid sample of $NH_4H_2PO_4$ at $B_0 = 4.7$ T with MAS spin rates 383 Hz (a), and 1245 Hz (b)

Source: Adapted with permission from K. Eichele and R.E. Wasylishen (1994), *J. Phys. Chem.* **98**, 3108–13. Copyright © 1994 American Chemical Society.

Eq. 7.3 In axial symmetry the shielding anisotropy is given by:

$$\Delta\sigma = \sigma_\parallel - \sigma_\perp$$

Eq. 7.4 The shielding anisotropy in arbitrary symmetry is given by:

$$\Delta\sigma = \sigma_{33} - \frac{1}{2}(\sigma_{11} + \sigma_{22})$$

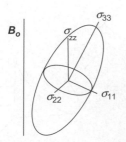

Figure 7.5 The principal axis system

The principal axis system

We can always define an axis system—called the principal axis system—in which all off-diagonal elements of the shielding tensor are zero (Figure 7.5). This allows us to describe the shielding for arbitrary symmetry with three quantities: σ_{11}, σ_{22}, and σ_{33}; by convention, $\sigma_{11} < \sigma_{22} < \sigma_{33}$. In cubic symmetry $\sigma_{11} = \sigma_{22} = \sigma_{33} = \sigma_{iso}$ and a sharp line is obtained.

If the molecule has axial symmetry (as in case of Ph_3PO molecule), we can define two important orientations and hence shifts: when the symmetry axis is parallel, $\sigma_\parallel$, or perpendicular, $\sigma_\perp$, to the magnetic field (Figure 7.2, equation 7.3). For an arbitrary asymmetric molecule, the lineshape of the signal due to chemical shift anisotropy will depend on the ratio between σ_{11}, σ_{22}, and σ_{33} (Figure 7.6, equation 7.4).

7.4 Dipolar interactions

The NMR spectrum is also affected by the magnetic moments of neighbouring nuclei through dipolar interactions. Relaxation effects were discussed in Chapter 4, section 4.4, and here we look at the effect on the NMR lineshape. Consider a pair of spin ½ nuclei: the magnetic moments can align either parallel or anti-parallel to each other, the moment of one increasing or decreasing the local field seen by the neighbour. This is a through-space coupling and distinct from the through-bond scalar coupling in solution NMR spectra discussed in previous chapters. Dipole–dipole coupling depends on the relative orientation and distance between the two spins; it is a tensor quantity.

The dipolar coupling constant d (Hz) between two spins I and S, depends on θ the angle between spins and, like the CSA, is averaged to zero in solution by molecular tumbling (averaging the $3cos^2\theta-1$ term to zero). In solid samples, where rapid molecular tumbling cannot occur, very large effects can be seen giving linewidths of tens to hundreds of kilohertz.

In a single crystal of a material containing isolated pairs of spins where there is a single value of θ, dipolar coupling will result in a doublet. In a powder sample, however, the grains of powder in the sample will be randomly oriented, so all possible values of θ occur and d will have a range of values reflecting the many orientations of the molecules/spins. This results in the characteristic lineshape known as a Pake doublet, which can be seen in Figure 7.7. This shows Pake's original solid-state ^{1}H NMR spectrum of isolated water molecules in gypsum powder ($CaSO_4\cdot2H_2O$). The Pake doublet is *ca.* 100 kHz wide and can be used to measure internuclear distance, since the splitting (in Hz) between the two maxima of the pattern is 3d/2.

Since dipolar coupling is a through-space effect, an alternative way to minimize dipolar interactions is to work with dilute spin systems such as ^{13}C at natural abundance where the chance of having two ^{13}C spins close together is remote or to replace some of the protons in the sample with deuterium; in effect this moves the proton spins further apart. However, this restricts the number of systems available to study, does nothing to remove the CSA, and can cause signal-to-noise problems—the sensitivity of a system of dilute spins is necessarily low.

Effect of magic angle spinning

Just as CSA can be removed by spinning the sample at the magic angle, so too can dipole–dipole coupling be reduced by MAS. However, for protons and for paramagnetic samples, extremely fast spinning rates—up to 110 kHz—may be required to resolve or eliminate the side band patterns (Figures 7.1, 7.4). At lower spinning rates, strong residual dipolar interactions give severe overlap of the resonances in ^1H spectra and broad, featureless lines in paramagnetic systems. The impact of spectral overlap due to ^1H–^1H dipolar interactions can be further reduced by using very high spectrometer fields to 'stretch' the chemical shift axis, since shielding increases linearly with the magnetic field strength while the dipolar interactions remain constant. Note, however, that dipolar interactions are always several orders of magnitude smaller than the value of the B_0 field.

An example of the combination of all three techniques—fast MAS, isotopic dilution, and very high field—is shown in Figure 7.8, which shows the ^1H NMR (21.1 T, ^1H = 900 MHz) spectrum of the metal-organic framework (MOF) α-$Mg_3(HCO_2)_6$. On dilution of the proton spins with deuterium, a shoulder begins to appear on the resonance on spinning at 18 kHz; compare Figures 7.8a and c. Spinning at 62.5 kHz allows the resonances of individual proton sites to be distinguished in the undiluted sample (figure 7.8b), and these become fully resolved on combining fast MAS with replacing 80% of the protons by deuterium (Figure 7.8d).

7.5 Polarization transfer in solid-state NMR

High-power decoupling is routinely used in multinuclear, solid-state NMR to eliminate dipolar coupling between (usually) proton spins and the observed nucleus. The removal of dipolar interactions decreases the linewidth, aiding sensitivity. However, since the aim is to remove the large dipolar couplings rather than the much smaller-scalar couplings as in the solution experiment, very high decoupling powers are needed. Fortunately, the high-power irradiation has another use; it can be used to enhance the sensitivity of solid-state NMR through a technique known as cross-polarization.

Cross-polarization

Many nuclei of interest in solid-state chemistry have low natural abundance compounding the problems of the inherently low sensitivity of NMR. Furthermore, T_1 relaxation times of spin ½ nuclei in the solid state are typically very long due to the restricted mobility (see Chapter 4, Figure 4.10) requiring long inter-scan delays if signal averaging is used to improve signal-to-noise, resulting in unfeasibly long experiment times (several days). Polarization transfer from neighbouring,

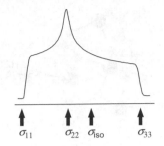

Figure 7.6 A schematic powder pattern caused by shielding anisotropy for arbitrary symmetry showing the positions of σ_{11}, σ_{22}, σ_{33} and σ_{iso}

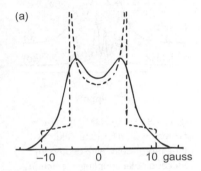

(a)

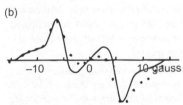

(b)

Figure 7.7 The solid-state NMR spectrum of H_2O in gypsum shows a Pake doublet, *ca.* 100 kHz wide. Pake's original spectrum, recorded in differential mode, is shown (b) and the calculated powder pattern and absorption line (a).

Source: Adapted with permission from G. E. Pake (1948), *J. Chem. Phys.*, **16**, 327. Copyright © 1948 American Institute of Physics.

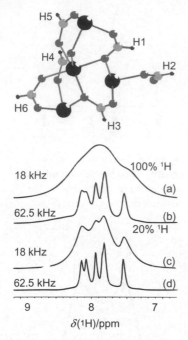

Figure 7.8 ^{1}H MAS NMR spectra of α-magnesium formate. Dilution of the proton spins with deuterium reduces dipolar coupling interactions, increasing the resolution. Dilution and fast-MAS resolves the different proton environments, H1–H6, right to left, in the spectra.

Source: Reprinted with permission from J. Xu et al. (2015), Chem. Mater., 27, 3306–16. Copyright 2015 American Chemical Society.

sensitive spins **I** of high magnetogyric ratio with fast spin–lattice relaxation rates to a less sensitive nucleus, **S**, can be used both to increase the polarization of the **S** spins and the scan repetition rate; in a sense, cross-polarization (CP) is the solid-state equivalent of the INEPT family of experiments in solution-state NMR. However, unlike the solution-state experiments that use scalar coupling to transfer polarization, in CP dipole–dipole interactions are used.

CP is achieved by first exciting the **I** spins and then arranging for the **I** and **S** spins to precess in their respective rotating frames at the same frequency—this is achieved by applying spin–lock fields to both the **I** and **S** spins such that $\gamma_I B_1^I = \gamma_S B_1^S$, where B_1^I and B_1^S are the spin–lock fields for the **I** and **S** spins (Figure 7.9). This is called Hartmann-Hahn matching and allows polarization transfer *via* the dipolar coupling from the **I** to the **S** spins; the maximum possible intensity at the **S** spins is γ_I/γ_S. The required inter-scan delay depends only on the relaxation rate of the faster relaxing **I** spins, so more scans can be acquired in the same time, boosting signal to noise and/or reducing the experiment time. Cross-polarization is now routinely used with MAS and heteronuclear decoupling for recording NMR spectra in the solid state. This introduces a wrinkle into the experiment: since the purpose of MAS is to remove dipolar coupling, how can polarization transfer occur? The answer is that dipolar coupling is not completely averaged to zero but becomes dependent on the MAS rotation frequency. We can reintroduce the required dipolar coupling during the spin–lock by setting $\gamma_I B_1^I = \gamma_S B_1^S +/- \omega_R$ where ω_R is the MAS rotation frequency.

Cross-polarization build-up

The drawback of CP is that spectra are no longer quantitative; the intensity of each resonance depends on the effectiveness of polarization transfer, which not only decreases with distance between the spins but also depends on the mobility of the spins, which affects relaxation rates. Cross-polarization builds up during the contact time when the irradiation used to transfer polarization between spins is 'switched on', increasing the **S** spin signal. However, relaxation of the **S** spins also occurs during this time, reducing the signal; signal intensity thus initially builds up due to CP and then falls away due to relaxation. The rate of decline will be higher for mobile spins which relax faster (Chapter 4); build-up curves can, therefore, be used to probe distances between spins and mobility. Two models, the *I–S* and *I–I*–S*, are commonly used to analyse CP kinetics: the models differ in that the latter considers the effect of *I* spins close to the *S* spins separately from *I* spins in the bulk. Which model best fits a given data set must be found by trial and error; a complete discussion is beyond the scope of this book, and interested readers are referred to the review in the Bibliography.

This is illustrated in the ^{13}C spectra of eggshell. Eggshell is a product of biomineralization and contains both inorganic, immobile calcite carbonate and organic components, mobile proteins, and lipids. The build curves from a series of ^{13}C CP MAS NMR spectra with variable contact time (Figures 7.10) were fit to the *I–I*–S* model (Figure 7.11), and allow the resonances of the two types of species present to be distinguished. Resonances in the region of 10–60 ppm build up quickly but decay rapidly, indicating that these are mobile species, i.e. proteins and lipids. The two carbonyl group signals can be easily differentiated by inspection of the CP kinetic fits (Figure 7.11), that at 172.3 ppm being due to a mobile species, and thus assigned to organic species, while that at 168.1 clearly being immobile, inorganic carbonate.

2-D polarization transfer experiments in the solid state

Metal-phosphine complexes are arguably the most potent catalysts of modern transition-metal catalysed reactions. While a variety of NMR methods are available to study their behaviour in solution, only recently has the combination of ^{1}H,^{31}P cross-polarization, MAS, and high-power proton decoupling become a routine tool for the characterization of metal-phosphine species in the solid state. ^{31}P CP/MAS allows the assignment of the chemical shifts and the determination of the coordination environment in metal-phosphine complexes.

The 1-D ^{31}P CP/MAS spectrum of a solid sample of the red polymorph of RhCl(PPh$_3$)$_3$ (Figure 7.12) reveals three isotropic shifts rather than the two expected from solution NMR; in the solid state, Wilkinson's catalyst adopts a distorted square planar structure in which the three phosphorus atoms are no longer equivalent. The ^{31}P NMR spectrum shows the **ABM** part of an ABMX spectrum (X = ^{103}Rh, I = spin ½; 100%). The doublet around 50 ppm can be assigned to P^1 *trans* to Cl based on the large ^1J(RhP) (*cis* coupling to P$^{2/3}$ is not resolved) (see Chapter 3). The resonances of P$^{2/3}$ show an ABX pattern in which both ^1J(RhP) and *trans*-$^2J_{P,P}$ are resolved and roofing is seen (see Chapter 2), the multpliets are doublets of doublets, not doublets of triplets.

Just as 2-D experiments can be used to highlight particular interactions in solution, so too 2-D experiments can be used in the solid state. However, there is a significant difference between the two regimes. In solution the scalar, or J coupling, is most frequently used for polarization transfer, whereas in the solid state dipolar interactions are used.

For example, solid-state NMR has been used to characterize a version of Wilkinson's catalyst in which the phosphine ligands is a triphenyl phosphine functionalized polymer grafted on to silica. First, a 2-D-^{31}P-^{1}H HETCOR spectrum using dipolar coupling, i.e. a cross-polarization CP step, was used to confirm incorporation of triphenylphosphine into the grafted polymer (Figure 7.13a). Note that because the experiment uses dipolar couplings, a through-space correlation is seen between phosphorus and the -CH-CH$_2$- resonances of the polymer chain. at *ca.* 1.5 ppm. A correlation between phosphorus and the aromatic proton resonances of the triphenylphosphine at *ca.* 6.9 ppm is also seen, confirming incorporation of the phosphine into the polymer.

Subsequent treatment with RhCl$_3$·xH$_2$O produced Wilkinson's catalyst-like structures in the polymer side chains. Scalar ^{31}P–^{31}P is seen in the 2-D J-resolved ^{31}P NMR spectrum of the catalyst (Figure 7.13b); the coupling constant ca. 400 Hz is consistent with the *trans*-^2J(PP) coupling observed in crystalline Wilkinson's

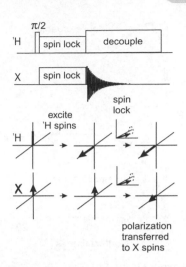

Figure 7.9 In cross-polarization, the sensitive high γ spins are excited, usually ^{1}H. Spin–lock fields are then applied to make the proton and X spins precess at the same frequency; resonance transfer of energy then occurs, resulting in observable magnetization of the X spins. N.B. The X spins are excited by transfer of energy from the proton spins, not by an rf pulse.

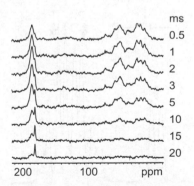

Figure 7.10 ^{13}C CP MAS NMR spectra of chicken eggshell recorded with different contact times (0.5–20 ms). Note the differences in signal intensity as contact time is varied.

Source: Adapted with permission from D. M. Pisklak et al. (2012), *J. Agric. Food Chem.*, **60**, 12254–9. Copyright © 2012 American Chemical Society.

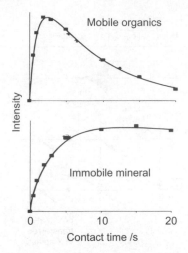

Figure 7.11 *I–I*–S* fits of the CP build-up curves for the carbonyl groups of the mobile (organic) and immobile (carbonate) in eggshell. Efficient relaxation causes the signal intensity to decline rapidly for mobile species.

Source: Adapted with permission from D. M. Pisklak et al. (2012), *J. Agric. Food Chem.,* **60**, 12254–9. Copyright © 2012 American Chemical Society.

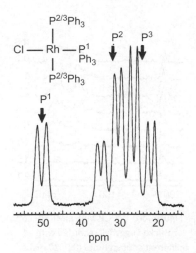

Figure 7.12 ^{31}P CP/MAS spectrum of the red polymorph of [RhCl(PPh$_3$)$_3$]

Source: Adapted with permission from G. Wu and R. E. Wasylishen (1992), *Organometallics,* **11**, 3242–8. Copyright © 1992 American Chemical Society.

catalyst above and suggests the rhodium is bound to the grafted polymer (Figure 7.13b). The weak signal at –10 ppm does not show ^{31}P–^{31}P coupling and was assigned to either monocoordinated or *cis*-coordinated phosphine groups for which *J*-coupling would not be resolved in the solid state, although the chemical shift, –10 ppm, is more consistent with excess phosphine groups not co-ordinated to the metal.

7.6 Quadrupolar interactions in solid-state NMR

The interaction of the quadrupole moment with the electric field gradient (EFG) present at the nucleus changes the separation of the nuclear spin energy levels (Figure 7.14), and causes enhanced relaxation of the nuclear spins. Several parameters are required to quantify/describe the NMR line of quadrupolar nuclei: the quadrupolar coupling constant (C_Q); the quadrupolar splitting parameter (ω_Q); the asymmetry parameter (η_Q); and the quadrupolar product (P_Q). These are usually determined by computer fitting of the experimental spectrum. C_Q describes the magnitude of the quadrupolar interaction, which can be up to 30 MHz in size, depending on the quadrupole moment and the electric field gradient at the nucleus. In cubic symmetry $C_Q = 0$; as the symmetry of the site decreases, so C_q increases, resulting in ever broader NMR lines. ω_Q is related to the change in energy of the nuclear spin energy levels produced by the quadrupolar interaction (Figure 7.14), while η_Q reflects the lineshape. The effect of first-order quadrupole interactions, which have a $3\cos^2\theta - 1$ dependence, can be reduced/removed by magic angle spinning at a sufficiently fast rate. However, not all quadrupolar interactions can be removed by magic angle spinning and much useful information about structure, site symmetry, mobility, etc. can be obtained by analysis of the quadrupolar lineshape.

Despite the significance of nickel compounds in heterogeneous catalysis and materials chemistry the first ^{61}Ni (I = ³⁄₂) solid-state NMR spectra of diamagnetic nickel complexes including [Ni(cod)$_2$], *bis*(1,5-cyclooctadiene) nickel(0), tetrakis(triphenylphosphite)nickel(0), [Ni[P(OPh)$_3$]$_4$], and tetrakis(t-riphenylphosphine)nickel(0) [Ni(PPh$_3$)$_4$] were reported only recently. Figure 7.15 compares the static ^{61}Ni NMR spectra of [Ni(cod)$_2$] and [Ni(PPh$_3$)$_4$]. DFT calculations of the lineshape for the cod compound reveal this to be dominated by the CSA, despite the large C_Q (2 MHz). The static ^{61}Ni solid-state NMR spectrum of Ni(PPh$_3$)$_4$, on the other hand, reveals a narrow (30 kHz) central transition with two 'horns' as expected for a quadrupolar I = ³⁄₂ nucleus, the central (m = ½ ↔ –½ transition) is flanked by satellite transitions (m = ³⁄₂ ↔ ½ and m = – ½ ↔ –³⁄₂). The intensity of the resonance is, however, unexpectedly low. This is attributed to the eqm between [Ni(PPh$_3$)$_4$] and [Ni(PPh$_3$)$_3$] + PPh$_3$ lying substantially to the side of the *tris*-PPh$_3$ compound in the solid state; the lower symmetry of the trigonal planar compound prevents observation of its NMR resonance due to extreme line broadening.

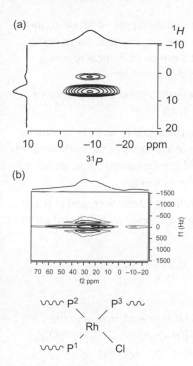

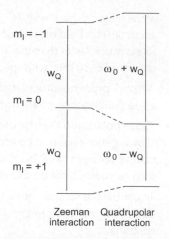

Zeeman   Quadrupolar
interaction   interaction

Figure 7.14 The quadrupolar interaction changes the nuclear spin energy levels so that these are no longer equally spaced. Shown here is the effect on a spin 1 nucleus. In a single crystal, two lines separated by $2\omega_Q$ will be observed.

Figure 7.13 (a) ^{31}P, ^{1}H HETCOR spectrum of the silica nanoparticles grafted with the PTPPE. F_1 is the ^{1}H dimension, F_2 is the ^{31}P dimension. Correlations between phosphorus and both the aliphatic proton resonances of the polymer chain and the aromatic proton resonances of the triphenylphosphine are seen, demonstrating that the phosphine was incorporated into the polymer. (b) 2-D J-resolved ^{31}P NMR of silica-immobilized Wilkinson's-type catalysts; the ^{2}J(PP) *trans*-coupling between P^1 and P^3 can be seen in F_1.

Source: Adapted with permission from S. Abdulhussaian et al. (2014), *Chem. Eur. J.*, **20**, 1159–66. Copyright © 2014, Wiley.

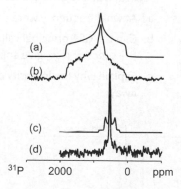

Figure 7.15 ^{61}Ni solid-state NMR spectra of [Ni(cod)$_2$] and [Ni(PPh$_3$)$_4$]

Source: Adapted with permission from P. Werhun and D. L. Bryce (2017), *Inorg. Chem.*, **56**, 9996–10006. Copyright © 2017 American Chemical Society.

7.7 **Summary**

- Slow motion of species in the solid state reintroduces dipole–dipole, chemical shift anisotropy, quadrupolar and other interactions that are averaged out in solution by molecular tumbling, resulting in broad NMR spectra.

- Dipole–dipole, chemical shift anisotropy, quadrupolar interactions have an angular dependence that contains the term $3\cos^2\theta - 1$. This term vanishes at the magic angle θ = 54.74°. Spinning the sample rapidly at 54.74° to the spectrometer field thus reduces these interactions, provided the spinning rate is much faster than the interaction to be removed. Typical spinning rates are 16–30 kHz, with speeds up to 125 kHz possible.

- Second-order quadrupolar and some other interactions have a different angle dependence, so are not removed by MAS.

- Cross-polarization can be used to improve the signal to noise in spectra by allowing more scans to be acquired for signal averaging in the same amount of time. Analysis of CP build-up rates provides information about things such as internuclear distances and molecular motions.

- In contrast to solution-state NMR, the dominant interaction used in polarization transfer experiments in the solid state is dipolar coupling, not scalar coupling.

7.8 Exercises

7.1. Figure 7.16 shows the ^{103}Rh{^{1}H} CP/MAS NMR spectrum of, and the CP build-up curve for, the dimeric Rh cation $[(H_2O)_4Rh(\mu^2\text{-}OH)_2Rh(H_2O)_4]$, recorded at two different MAS rates (1.0 kHz, top; 4.0 kHz).

   a. At what frequency were the spectra recorded?

   b. Can the CSA principal values of δ_{11}, δ_{22}, δ_{33}, and δ_{iso} be estimated from the spectra? If so, give approximate values.

   c. Explain why the intensity of the CP spectrum first increases, then dies away.

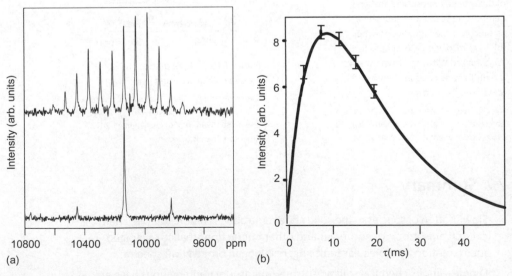

Figure 7.16 (a) ^{103}Rh{^{1}H} CP/MAS NMR spectrum of; and (b) CP build-up curve for, $[(H_2O)_4Rh(\mu^2\text{-}OH)_2Rh(H_2O)_4]$

Source: Adapted with permission from B. L. Phillips et al. (2006), *J. Amer. Chem. Soc.*, **128**, 3912–13. Copyright © 2006 American Chemical Society.

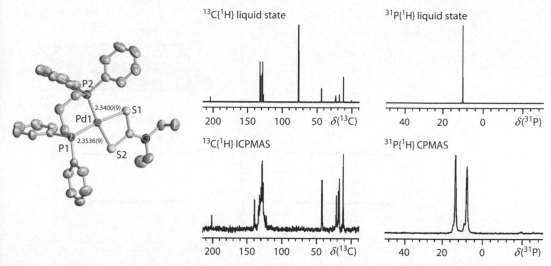

Figure 7.17 $^{13}C\{^1H\}$ and $^{31}P\{^1H\}$ spectra of [Pd(dppp)(dtc)]BF$_4$ (dppp = diphenylphosphinopropane; dtc = *N,N*-diethyldithiocarbamate) in solution and anchored to a silica surface

Source: Adapted with permission from J. W. Wiench et al. (2009), *J. Amer. Chem. Soc.*, **131**, 11801–10. Copyright © 2009 American Chemical Society.

7.2 Figure 7.17 shows the $^{13}C\{^1H\}$ and $^{31}P\{^1H\}$ spectra of [Pd(dppp)(dtc)]BF$_4$ (dppp = diphenylphosphinopropane, dtc = *N,N*-diethyldithiocarbamate) in solution and anchored to a silica surface.

   a. Explain the differences between the solution and solid-state spectra.

   b. What effect often results in additional resonances being seen in the solid-state NMR spectrum of a crystalline sample?

7.3 In a cross-polarization experiment, a 9.4 T spectrometer was used; ^1H was the polarization source and ^{109}Ag was the polarization receiver.

   a. What is the maximum signal enhancement that can be obtained neglecting relaxation effects?

   b. The proton B_1 field used was 1.4 mT. What B_1 field is required at ^{109}Ag to fulfil Hartmann-Hahn matching condition at a MAS rate of 16 kHz?

7.4 The cathode (positive electrode) of a rechargeable battery is the electrode where reduction takes place during the discharge cycle; for lithium-ion cells this is the lithium-based electrode. Li-ion batteries typically comprise a lithium-doped cobalt oxide (LCO) positive electrode, a graphite negative electrode, a porous separator, and a solution of a lithium salt in an organic liquid as electrolyte (Figure 7.18). Overcharging of lithium-ion batteries is reported to promote the growth of lithium-metal dendrites on the cathode. Metal conduction electrons shield the metal nuclear spins in metals which characteristically show large down-field shifts (the Knight shift), whereas Li-ions in solution or in insulators typically have very small chemical shifts.

   a. Assign the ^7Li solid-state NMR spectra shown in Figure 7.18. The ^7Li resonance of LCO is very broad and the electrolyte used is 1 M LiPF$_6$.

   b. Is your assignment consistent with the formation of Li dendrites?

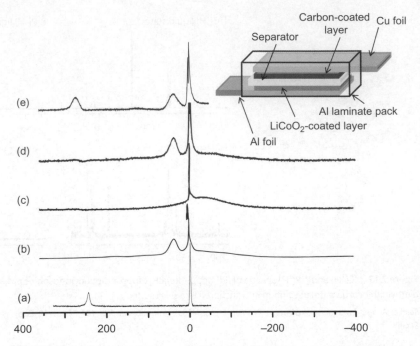

Figure 7.18 ^{7}Li solid-state NMR spectra of lithium ion cells. (a) Li metal with electrolyte (LiPF$_6$); (b–c) LiCoO$_2$/electrolyte/graphite cell during charge–discharge–recharge; (e) cell after overcharging. In the charged state, Li is stored as lithium carbides.

Source: Adapted with permission from J. Arai et al. (2015), *J. Electrochem. Soc.*, **162**, A952–A958. Copyright © 2015 The Electrochemical Society.

Glossary

AB, ABX, ABMX, etc Nomenclature for spin systems. Closely similar groups of spins are assigned letters close to each other in the alphabet, dissimilar groups well separated letters. Primes are used to indicate chemically equivalent but magnetically inequivalent groups, e.g. AA′, XX′. etc.

Chemical shift Dimensionless quantity used to report the position of a multiplet in the NMR spectrum with respect to a reference.

Chemical shift anisotropy Dependence of the chemical shift on the orientation of a molecule with respect to B_0. Important in the solid state but normally averaged out by fast molecular motions in solution.

Chemical shielding Screening of the nuclear spin from the spectrometer magnetic field by core and bonding electrons. Gives rise to the chemical shift.

COSY A homonuclear 2-D experiment that correlates groups that are scalar coupled to each other.

Dynamic process See Fluxional and Exchange processes.

Exchange limit–fast Dynamic processes are occurring at a rate above the coalescence temperature. An averaged signal is observed.

Exchange limit–slow Dynamic processes are occurring at a rate below the coalescence temperature. Separate signals are observed for each site in the dynamic system.

Exchange process An intermolecular process that moves groups from one molecule to another.

EXSY A homonuclear 2-D experiment that correlates groups that are involved in a dynamic process.

First-order spectrum NMR spectrum which obeys 'simple' scalar coupling rules due to high symmetry and/or well-separated chemical shifts.

Fluxional process An intramolecular process that moves groups in a molecule from one site to another or promotes a change in conformation.

Free induction decay The decaying signal induced in the NMR coil as the magnetization precesses about B_0 following an excitation pulse.

Gyromagnetic ratio The proportionality constant between the nuclear spin, I, and the nuclear magnetic moment, μ.

Heteronuclear Spins of a different element or isotope of an element.

High-resolution spectrum NMR spectrum recorded in solution. Molecular tumbling results in sharp lines for spin ½ nuclei.

HMBC A heteronuclear 2-D experiment that correlates groups that are scalar coupled to each other over multiple bonds.

HMQC A heteronuclear 2-D experiment that correlates groups that are scalar coupled to each other using multiple quantum coherences.

Homonuclear Spins of the same element or isotope of an element.

HSQC A heteronuclear 2-D experiment that correlates groups that are scalar coupled to each other using single quantum coherences.

Intermolecular exchange Exchange of a substituent between molecules.

Intramolecular exchange Exchange of a substituent between sites within a molecule; a fluxional process.

Isotope Nuclides of the same element having different numbers of neutrons. Can have a different spin and will have a different resonance (Larmor) frequency.

Isotopologue A molecule having the same structure and isotopic composition but a different placement of isotopes within the structure.

Larmor frequency The NMR resonance frequency of the spin being observed.

Magic angle spinning A solid-state NMR experiment that mechanically spins the sample at kHz speeds at 54.74° with respect to the static magnetic field to average out line broadening effects such as chemical shift anisotropy, dipolar coupling, etc.

Magnetic inequivalence Occurs when the spins in one chemically equivalent group couple differently to the spins of another group of spins.

NMR timescale Has various meanings; the time for which a molecule must exist in a given state for its NMR spectrum to be observed.

NOE, NOESY, HOESY 1-D and 2-D experiments that correlate groups that are close to each other in space but not directly bonded to each other.

Nuclear spin The property of the nucleus that underlies NMR spectroscopy.

Quadrupolar nucleus A nucleus with spin greater than ½. Normally gives fast relaxation and broad NMR lines.

Relaxation The processes by which an excited system returns to equilibrium.

Relaxation–spin-lattice Relaxation due to loss of energy to the surroundings.

Relaxation–spin-spin Relaxation caused by transfer of energy between spins.

Satellites Weaker resonances that flank the main resonance due to coupling to a nuclide that has low natural abundance. The sub-spectrum due to such coupling.

Second-order spectrum NMR spectrum which does not obey 'simple' scalar coupling rules due to small chemical shift differences between coupled groups or magnetic inequivalence.

Solid-state NMR NMR spectra recorded in the solid state give broad lines due to effects such as chemical shift anisotropy, dipolar coupling, etc. that are not averaged to zero by molecular tumbling.

Vector model A classical description of the NMR experiment used to explain simple NMR experiments and relaxation.

Bibliography

Abragam, A. (1978), *The Principles of Nuclear Magnetism*, Clarendon Press, Oxford.
The NMR theorist's 'bible'—not for the faint hearted!

Brevard, C. and Granger, P. (1981), *Handbook of High-Resolution Multinuclear NMR*, Wiley Interscience, Chichester.
A valuable compilation of NMR data.

Derome, A. E. (1987), *Modern NMR Techniques for Chemistry Research*, Pergamon, Oxford.
An excellent reference texts dealing with all aspects of NMR spectroscopy, with good sections on multiple pulse and 2-D methods.

Duer, M. J. (2005), *Introduction to Solid-State NMR Spectroscopy*, Oxford, Blackwell Publishing Ltd.

Freeman, R. (1997), *Spin Choreography*, Oxford, Oxford University Press.

Fyfe, C. A. (1987), *Solid State NMR for Chemists*, C.F.C. Press, Guelph, Ontario, Canada. An introduction to NMR in the solid state.

Harris, R. K. and Mann, B. E. (1978), *NMR and the Periodic Table*, Academic Press, London.

Hoare, P. J. (2015a), *NMR: The Toolkit, 2015*, 2nd edn, Oxford, Oxford University Press.
An introduction to pulse sequences used in liquid-state NMR—a companion volume to this text.

Hoare, P. J. (2015b), *Nuclear Magnetic Resonance*, 2nd edn, Oxford, Oxford University Press.
An accessible introduction to the physical principles of liquid-state NMR—a companion volume to this text.

Keeler, J. (2004) http://www-keeler.ch.cam.ac.uk/lectures/understanding/chapter_3.pdf.
Gives a complete mathematical description of pulses in NMR spectroscopy.

Keeler, J. (2010), *Understanding NMR Spectroscopy*, 2nd edn, Oxford, Wiley.
A clear and accessible exposition of the physical basis and concepts in modern NMR spectroscopy aimed at people who have some familiarity with high-resolution NMR and who wish to deepen their understanding of how NMR experiments actually 'work'.

Kolodziejski, W. and Klinowski, J. (2002), 'Kinetics of Cross-Polarization in Solid-State NMR: A Guide for Chemists', *Chem. Rev.* **102**, 613–28.
A guide to CP kinetics.

Laszlo, P. (ed.) (1983), *NMR of Newly Accessible Nuclei*, Academic Press, New York.

Mason, J. (ed.) (1987), *Multinuclear NMR*, Springer, New York.
Two invaluable reference texts for the inorganic chemist.

Index

Note: Tables and figures are indicated by an italic *t* and *f* following the page number.